Y0-BSL-828

Petroleum Engineering

Geology of Petroleum

Edited by Heinz Beckmann

Vol. 3

PROPERTY OF
CHEVRON OIL FIELD RESEARCH CO.
LA HABRA, CALIF.

Petroleum Engineering

by Alfred Mayer-Gürr

CHEVRON OIL FIELD RESEARCH COMPANY
PROJECT COPY
APR 30 1981
La Habra, California

Ferdinand Enke Publishers Stuttgart 1976

Author:

Prof. Dr. *Alfred Mayer-Gürr*
Gewerkschaften Brigitta/Elwerath
D-3000 Hannover

CIP-Kurztitelaufnahme der Deutschen Bibliothek

Mayer-Gürr , Alfred
Petroleum engineering.
(Geology of petroleum ; Vol. 3)
ISBN 3-432-87681-5

ISBN 0-470-15082-3 John Wiley & Sons, New York — Toronto
ISBN 0-273-00960-5 Pitman Publishing, London
ISBN 91-85342-18-1 Scientia Bokförlag, Uppsala

With agreement of the Georg Thieme Publishing Company, Stuttgart, this book is published under the mark: flexi-book

All rights, including the rights of publication, distribution and sales, as well as the right to translation, are reserved. No part of this work covered by the copyrights hereon may be reproduced or copied in any form or by any means – graphic, electronic or mechanical including photocopying, recording, taping, or information and retrieval systems – without written permission of the publisher.

© Ferdinand Enke Verlag, Stuttgart 1976

Printed in Germany by Printing House Dörr (Adam Götz, proprietor), Ludwigsburg

For Béla Binder, Budapest

Preface

This little book has not been written for specialists in petroleum engineering. It is addressed to all the friends and colleagues in the oil and gas business, who feel that they would wish to look "over the fence" of their particular sphere of activity, to young students and all the others who are interested in the question: what is petroleum engineering? Not least to the practical men in oil and gas operations, who would like to know a bit more about the background of their work without having to bother with too much scientific theory.

This concept and the prescribed scope of the booklet are responsible for the great emphasis given to diagrams. In my opinion, a technical concept is most appropriately presented by a summarizing and explanatory schematic drawing. This is a careful and reliable simplification; it is an abstraction which contains the essential and leaves out all that is unessential.

Above all, however, this book is dedicated to the many friends and colleagues to whom the writer owes much of his knowledge through daily talk and discussion and whom he wishes to thank in this way. To these gentlemen the booklet owes much of its clarity, but none of its defects, for which they are not in the least responsible. It is a pleasure and an honour for me to mention in particular: Dr. *D. Betz*, Dr. *E. Bradel*, Dr. *H.-J. Drong*, Mr. *W. Friedrich*, Mr. *D. Fuhrberg*, Dr. *H. Leicht*, Dr. *D. Marsal*, Dr. *G. Mießner* and Dr. *F. Springer*.

For their active cooperation I am equally indebted to Miss *B. Boss*, Mr. *W. Nagorni*, Miss *T.-L. Rummel* and Mr. *F. Suhr*.

To the shareholders of my company I am – last but not least – indebted for their continued good favour.

All readers whose mother tongue is English should bear in mind how great is the language barrier not only in writing but also in thought.

Mayer-Gürr

Contents

Preface VII
1 *What is Petroleum Engineering?* 1
2 *The Reservoir* 3
2.1 What is a Reservoir? 3
2.2 Reservoir Rocks and their Properties 3
2.2.1 General 4
2.2.2 Porosity 8
2.2.2.1 Classification 9
2.2.2.2 Definition 10
2.2.2.3 Porosity of some Rocks 10
2.2.2.4 Measurement of Porosity 11
2.2.2.5 Pore Compressibility (Soil Subsidence) 14
2.2.3 Permeability 19
2.2.3.1 Darcy's Experiment 19
2.2.3.2 Definition 19
2.2.3.3 Measurement of Permeability 21
2.2.4 Other Parameters 23
2.2.5 Relationship between the Individual Parameters . . 25
2.2.6 Electrical Resistivity of Sandstones 27
2.3 The Content of the Reservoir 32
2.3.1 Saturation 32
2.3.1.1 Measurement on Core Samples 33
2.3.1.2 Capillary Pressure P_c as a Help to Determining Saturation 34
2.3.1.3 Saturation and Capillary Pressure 38
2.3.1.4 More on P_c-Curves 41
2.3.1.5 Determination of S_w, S_o, S_g from P_c-Curves . . . 42
2.3.1.6 Oil Migration and Capillary Pressure 44
2.3.2 Oil and Natural Gas under Reservoir Conditions (PVT Relationships) 46
2.3.2.1 Pressure Volume Relationship of Crude Oil 47
2.3.2.2 Pressure Temperature Relationship (Phase Relations) . 50
2.3.2.3 Viscosity of Crude Oil 52
2.3.2.4 Super Compressibility of Natural Gas 53
2.3.2.5 Sampling 56
2.4 Reservoir Pressure and Reservoir Temperature . . . 58
2.5 Reservoir Energy (The driving forces) 68
2.5.1 Dissolved Gas Drive 68
2.5.2 Gas Cap Drive 70
2.5.3 Water Drive 72
2.5.4 Other forces and combinations of several forces . . . 74
3 *Flow of Fluids through the Reservoir* 77
3.1 One Phase Flow 79
3.2 Multiple Phase Parallel Flow 81
3.2.1 Relative (and effective) Permeability 83

3.2.2 Mobility and f-Functions 86
3.3 Displacement of Oil (and Gas) by Water 93
3.3.1 Displacement Theories 94
3.4 Remarks on the Recovery Factor 102
3.5 Material Balance 107
3.6 Reservoir Models and Simulations 112
3.6.1 Reservoir Models 112
3.6.2 Mathematical Reservoir Simulation 114
4 *The Well – our Point of Observation* 119
4.1 Pressure Measurements 119
4.1.1 Methods and Devices 119
4.1.2 Capacity of a Well 121
4.1.2.1 The Productivity Index (PI) 121
4.1.2.2 Testing the Capacity of Gas Wells 124
4.1.3 Pressure Build-up Tests 127
4.1.3.1 General 127
4.1.3.2 Special Cases 133
4.1.4 Interference and Pulse Testing 137
4.1.5 Flow Tests 140
4.1.5.1 Reservoir Limit Tests 142
4.2 Production Control Measurements 143
5 *The Development of Oil- and Gas-Fields* 147
5.1 Estimation of Reserves 147
5.1.1 Volumetric Method 150
5.1.1.1 Recovery Factor 153
5.1.1.2 Classification of Reserves 154
5.1.1.3 Expectation Curves 154
5.1.2 "Dynamic" Methods 162
5.2 Some Considerations on Economics 164
5.3 The Development of Oil-fields 168
5.4 The Development of Gas-fields 171

Appendix A: How to determine average porosity 178
B: Methods for determining porosity 181
C: Determination of porosity on thin sections . . 182
D: X-ray diffraction analysis 183
E: Notes on capillarity 184
F: Hg-injection method to determine P_c . . . 186
G: Natural gas formation volume factor B_g . . . 187
H: Klinkenberg effect 188
I: Swelling phenomena 190
K: Frontal-advance-rate formula 192
L: Example of a simple material balance and oil-in-place calculation for an undersaturated oil-field which still produces above the bubble point . . 194
M: Example of an isochronal test in a gas well . . 195

N: Interpretation of a pressure build-up measurement in an oil well 196
O: Flowing pressure at different production rates . 198
P: Determination of the original "oil in place" by material balance in an undersaturated oil reservoir without water drive 200

References 203
Register 205

Symbols

A	=	area
a	=	air
	=	total expansion factor (oil + gas + water + rock) under reservoir conditions expressed in fractions of N
B	=	formation volume factor
B_o	=	oil formation volume factor
B_g	=	gas formation volume factor
B_w	=	water formation volume factor
B_t	=	total (two phase) formation volume factor
b_o	=	$\left(\frac{1}{B_o}\right)$ reciprocal oil formation volume factor (shrinkage factor)
BT	=	water break through
C	=	coefficient of gas-well back-pressure curve
c	=	compressibility (c_o = oil, c_w = water, c_b = rock [bulk], c_p = pore)
c_f	=	effective pore compressibility
D	=	depth
d	=	diameter
$\overline{d}_p$	=	diameter, mean particle
F	=	formation resistivity factor (R_o/R_w)
f	=	porosity factor
	=	fraction of a flow stream consisting of a particular phase
f_o	=	fraction of oil (f_w of water) in a flow stream
G	=	total initial gas in place
G_p	=	cumulative gas produced
g	=	acceleration of gravity
	=	gas
h	=	thickness
	=	height
I	=	resistivity index (R_t/R_o)
k	=	absolute permeability
k_{rg}	=	relative permeability to gas (k_{ro} = to oil, k_{rw} = to water)
L	=	length
M	=	mobility ratio (generally λ displacing/λ displaced medium
m	=	porosity (cementation) exponent
	=	ratio of initial free-gas volume to initial reservoir oil volume
	=	slope
N	=	initial oil in place
N_p	=	cumulative oil produced
N_{Re}	=	Reynold's number
N_{Rp}	=	pseudo Reynold's number
n	=	saturation exponent
	=	exponent of gas-well capacity test

	=	total moles
o	=	oil
OWC.	=	oil/water contact
P_c	=	capillary pressure
P_{cd}	=	displacement pressure
PI	=	productivity index
PI_s	=	specefic productivity index
p	=	pressure
p^*	=	extrapolated shut-in pressure (Horner)
p_b	=	bubble point (saturation) pressure
p_e	=	external boundary pressure
p_i	=	initial pressure
p_n	=	reservoir pressure at G_p
p_o	=	original pressure
p_s	=	static (shut-in) pressure
$p_{\Delta t}$	=	pressure after shut-in time Δt
p_w	=	bottom hole pressure (general)
p_{wf}	=	bottom hole pressure, flowing
p_{ws}	=	bottom hole pressure, static
q	=	production rate or flow rate
q_{tot}	=	total flow rate
R	=	universal gas constant (per mole)
R_o	=	resistivity of a formation 100 % saturated with water of resistivity R_w
R_p	=	cumulative gas-oil ratio
R_{si}	=	initial solution gas-oil ratio
R_w	=	resistivity of water
r	=	radius
r_e	=	external boundary radius
S	=	saturation
	=	skin factor
	=	specific inner surface area
$\overline{S}$	=	inner surface area
S_g	=	gas saturation (S_o = oil saturation, S_w = water saturation)
S_{gc}	=	critical gas saturation
S_{or}	=	residual oil saturation
S_{wi}	=	irreducible water saturation
S_{wr}	=	residual water saturation
s	=	skin effect
sc	=	standard conditions (as subscript)
T	=	temperature
	=	tortuorsity
T_s	=	average anual surface temperature
t	=	time
	=	tube
Δt	=	time after well is shut in

t_p = (pseudo time) equivalent time the well was on production prior to shut in
V = volume
V_b = bulk ($_m$ = matrix, p = pore) volume
V_n = volume under standard conditions
W = initial connate water
W_e = cumulative water influx (encroachment)
W_p = cumulative water produced
W_{pc} = cumulative connate water produced
x = distance (in Buckley-Leverett formula)
z = gas deviation factor (compressibility factor)
z_i = initial real gas factor (at p_i)
z_n = real gas factor (at p_n)
μ = viscosity
ϱ = density
θ = angle of contact
ϕ = porosity
σ = surface tension
λ = mean free molecular path
Δ = difference
λ = mobility (k/μ)

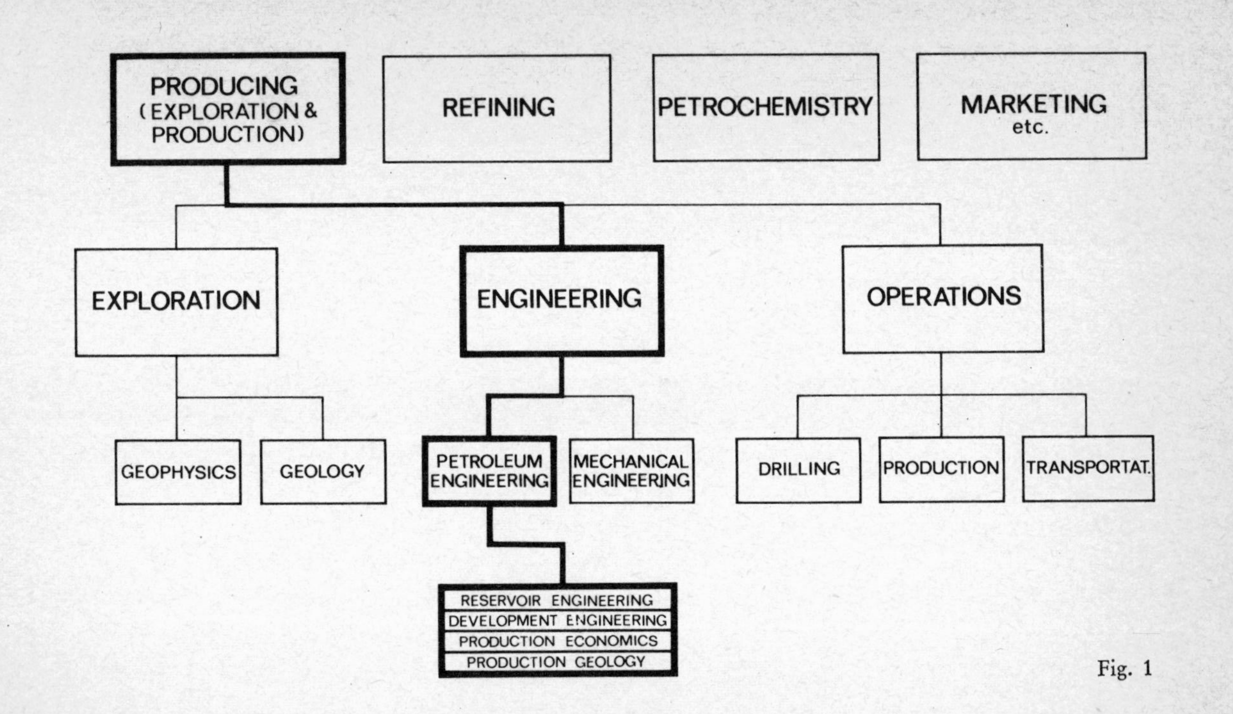
PRODUCING
(EXPLORATION &
PRODUCTION)
REFINING
PETROCHEMISTRY
MARKETING
etc.
EXPLORATION
ENGINEERING
OPERATIONS
GEOPHYSICS
GEOLOGY
PETROLEUM
ENGINEERING
MECHANICAL
ENGINEERING
DRILLING
PRODUCTION
TRANSPORTAT.
RESERVOIR ENGINEERING
DEVELOPMENT ENGINEERING
PRODUCTION ECONOMICS
PRODUCTION GEOLOGY

Fig. 1

1 What is Petroleum Engineering?

Exploration for and production of crude oil and natural gas (E & P) is only one branch of the oil and gas business as a whole; other branches are refining, distribution, petrochemistry etc. However, E & P is indeed the branch that stands at the very beginning, both historically and in the order of procedure.
The functions of EP (often called producing, in contrast to refining, marketing, etc.) can be grouped logically/organically and in operational/organizational terms in the following manner:

Exploration
Acquisition and maintenance of concessions,
geological and geophysical work,
locating of exploration wells.

Engineering
Following the discovery of a new field, this field has to be drilled up or "developed"; the development plan is based on accurate knowledge of the geological conditions (production geology), the reservoir (reservoir rock + its content), the behaviour of this reservoir under changing conditions (pressure, temperature, phase) and the economic aspects (production economics). For oil-fields, the development plan takes into account the reservoir parameters and the economic requirements; the development of gas-fields is more complicated, because it depends also on the requirements of the sales contract.
In addition the entire technical equipment of the field has to be planned, which is done by the engineers in the mechanical engineering department.

Operations
The drilling function (all kinds of wells: exploration, extension, production, capacity, auxiliary wells) and the production operations (including pipeline transportation) are the responsibility of the operations department.

Summarizing
Petroleum Engineering is the science (or the art) of planning the development and the production of oil- and gas-fields in such a way that an optimal recovery of oil and gas is achieved with optimal economic results.

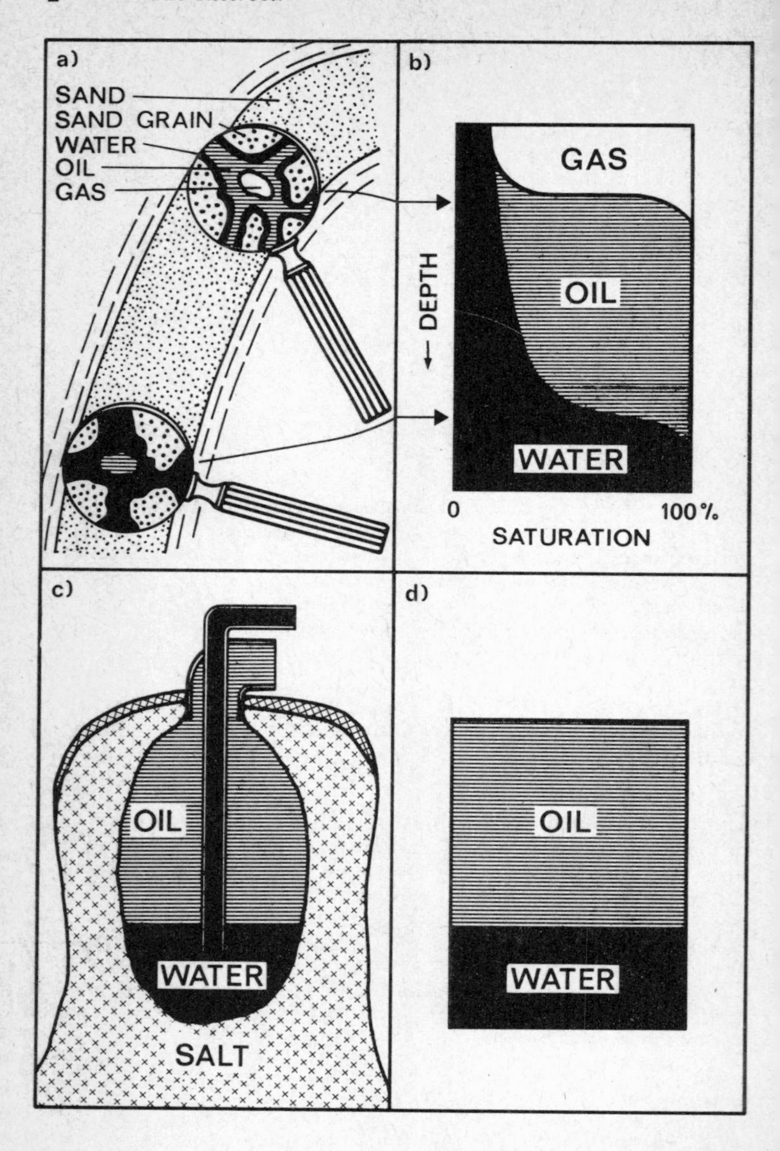
a)
SAND
SAND GRAIN
WATER
OIL
GAS
b)
GAS
OIL
WATER
DEPTH
0
100%
SATURATION
c)
OIL
WATER
SALT
d)
OIL
WATER

Fig. 2

2 The Reservoir

2.1 What is a Reservoir?

A reservoir (in the proper sense, as opposed to a mere accumulation) is a product of both reservoir rock and fluid content. Since the pores in most cases are of capillary size, they are governed by the laws of capillary action. Therefore, the distribution of gas, oil and water (saturation S_g, S_o, S_w) is dependent not only on the respective specific gravities of the fluids involved, but also on capillary action (capillarity). Along with capillarity, wettability is also an important factor. The distribution of gas, oil and water changes with variations in height over the free water table, namely when gravity and capillarity have reached equilibrium. Therefore, in a reservoir there is no sharp horizontal dividing line with gas or oil above and water beneath.
The difference between a reservoir (in the proper sense) and a mere accumulation becomes obvious when comparing a reservoir with an artificially constructed salt cavern filled with crude oil (Fig. 2):

a) Here is a cross-section of an anticline occurring in a porous sandstone payrock. If one were to examine a structurally high-lying layer, the surface of the sand grains is wetted with a water film, but the largest portion of the pore space is filled with oil and bubbles of gas. In a structurally deep-lying payrock layer, the pores are largely filled with water, and only within the pore-water are there small suspended drops of oil.

b) If one presents graphically the water, oil and gas saturation as it varies vertically, the lowest portion of a reservoir is 100 % water saturated. When dealing with a higher structural position, the percentage of water in the reservoir quickly drops off, while the oil percentage increases. But even at the highest point within the reservoir, water is present, filling approximately 10 % of the pore space.

c) In an artificial salt cavern leached out for the storage of oil, there is no effective capillary pressure; only gravity is operative. There is a sharp line of separation between oil and water, governed wholly by specific gravities.

d) In this illustration the vertical distribution of oil and water is shown in a mere accumulation where only gravity is effective. The difference, when compared with a real reservoir [as in illustration b)], is strikingly clear.

2.2 Reservoir Rocks and their Properties

For the petroleum engineer the most important payrock parameters are

its porosity
its permeability
its capillary pressure curve.
In addition, there are still other specifications for reservoir rocks.

2.2.1 General

From a geological point of view the following is of interest to the petroleum engineer: most reservoir rocks are formed by either
a) silicoclastic sediments, or
b) carbonate rocks.

a) The reservoir rocks formed by silicoclastic sediments which actually are reworked products of crystalline rocks or older compacted sediments, are best represented by the sandstone in the "sand" triangle of Fig. 3. Their particles consist mainly of quartz with varying amounts of feldspar ("arkose") and rock fragments as well ("graywacke"). Carbonate rock fragments occur mainly in the flysch or molasse sediments, which are made up of the erosional debris from mountain fold belts. (In such molasse sedi-

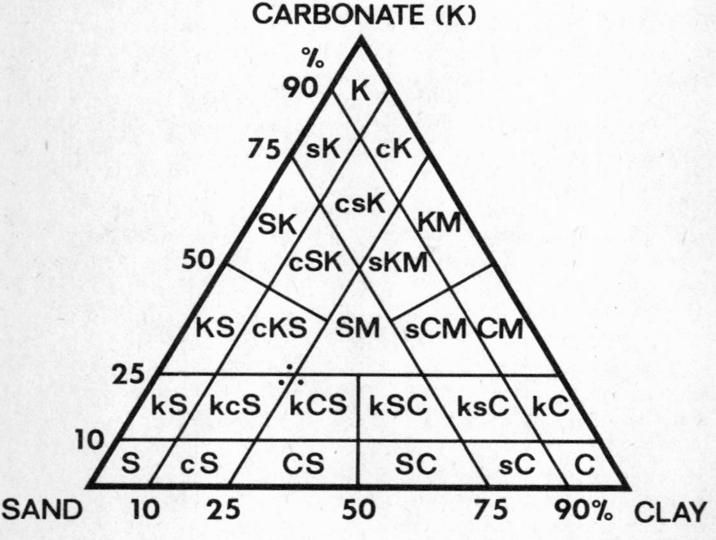

Fig. 3 Triangle for classification of sand-clay-carbonate rocks (After *Füchtbauer*[9]).

S = sand
C = clay
K = carbonate
M = marl

s = sandy
c = clay
k = calcereous
∵ = "marly sandstone"

ments, clay is frequently sedimented along with sand grains, especially in finegrained sands.) Clay always has a negative effect on reservoir quality.
Movement and deposition of sands may occur in various environments; (although most movement and deposition occurs in water) e.g.
in the deltaic estuaries of large rivers, or
along oceanic continental slopes ("turbidites", California), or
as barren sands along shallow coasts (in the Jurassic of NW-Germany), or
in terrestrial environments in rivers, as detritus fans, or even as dunes formed by winds (Groningen).

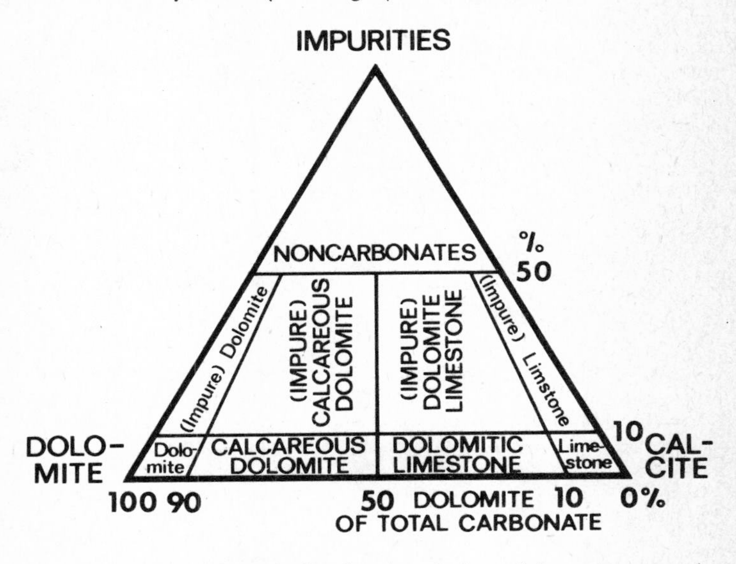

Fig. 4 Compositional graph showing the relations of the carbonate rocks. The impurities commonly consist of sands, clays, and shales. (After Leighton and Pendexter). (After *Levorsen*[15])

b) Reservoir rocks consisting of limestone or dolomite (Fig. 4) are also mostly of sedimentary origin. They are, however, mainly formed from organic debris. But there are also organically grown rocks, e.g. certain reefs along the shores of tropical seas, whose widenetted framework is filled with debris from the surrounding area and from the reef itself. Reefs consisting of algal crusts may grow up without any contribution by sediments. Because of the presence of organic elements, carbonatic reservoirs therefore con-

tain a fairly wide variety of rock types. The large group of "fossil limestones" is made up of extremely varied forms, each form depending largely on the type of organisms which have supplied the organic debris. "Oncolites", formed by algal crusts or "stromatolites" belong to this group.
Of the carbonates, those most closely related to the sandstones are the "carbonate-arenites", whose grains originate from the reworking of carbonate rocks. Pellet limestones (fecal pellets) and oolites (chemical precipitates) also generally show the clear texture of sediment particles and pore space between them like sandstones (Fig. 5).

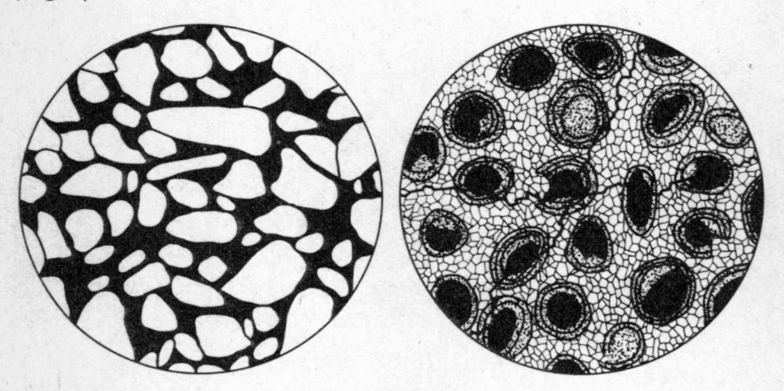

Fig. 5 Porosity of sandstones is, in general, "intergranular" (left figure shows a quartz sandstone; the pore volume is shown in black).
The right figure shows a dolomite-oncolite with "intragranular" pores and conveying canals (e.g. German "Hauptdolomit" of Upper Permian "Zechstein"). (By courtesy of Dr. *H.-J. Drong*, Brigitta/Elwerath, Hannover)

After their deposition, sediments are subjected to physical and chemical changes, which are summarized under the term *Diagenesis*.
Mechanically (compaction)
The beginning of this development is known as *compaction*. Compaction means a decrease in thickness[20] and this decrease is always combined with a reduction of pore space. An increasing overburden pressure results in a tighter packing of the grains and in the expulsion of pore water (upward-directed "compaction flow").
Chemically
If the maximum mechanical packing density has been reached, compaction is generally continued by a chemical process: the grains are partly dissolved at the tiny contact points upon which the petrostatic pressure is acting and they merge more and more into

one another. In general, the dissolved quartz crystallizes again into the pores ("authigenic quartz"), resulting in a (particularly marked) reduction in pore space.
Decreasing porosities in sandstones as a function of age and overburden can be shown in "compaction curves" (see Fig. 6). Other crystallization processes eventually contribute to lithification; for example, if feldspar were to be transformed into clay minerals (e.g. kaolinite, illite). This might result in a complete loss of permeability within the reservoir rock.

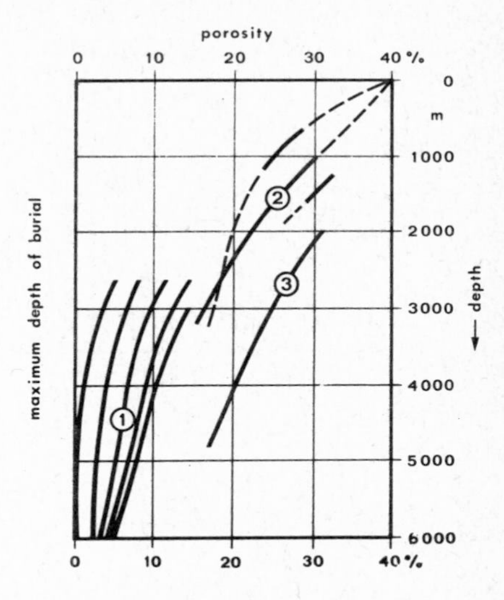

Fig. 6 Decrease of sandstone porosity with increasing maximum burial depth for Carboniferous micaceous silty sandstones (1) and for quartz sandstones of Jurassic-Cretaceous (2) and Tertiary age (3). (After *Füchtbauer*[9])

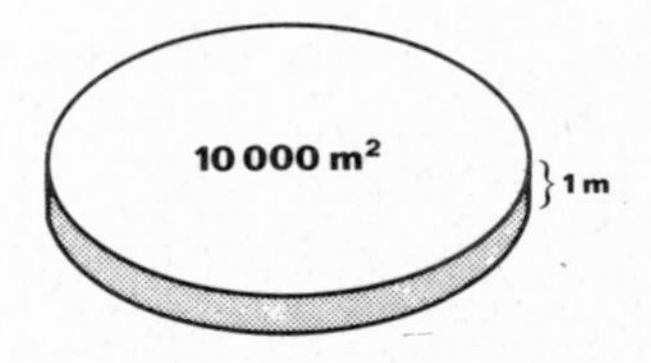

Fig. 7 1 ha/m, that is a reservoir rock with a productive area of 10,000 m² (one hektar = 1 ha) and with a thickness of 1 m contains at a porosity of 25 % (ϕ 25 % or f = 0.25) a pore volume of 2,500 m³. (Respective terms in U.S.A. are bbls per acre/foot)

From the pore water, carbonates or sulphates are precipitated as cement minerals and may displace other minerals. On the other hand, solution processes can result in the formation of new pores. Carbonate rocks particularly are subjected to manifold changes by recrystallization. In chemical terms, carbonates are much less stable than silicates, especially because of their sensitivity to CO_2; secondary pores often suffer from a lack of conveying channels.
The organic content of the sediments is particularly sensitive to diagenetic processes, as is generally known from the various stages of coal formation. The relatively small conversion steps are used in oil well sampling as a scale for diagenetic changes ("coalification"). Last but not least, the formation of hydrocarbons itself is a result of diagenesis.

2.2.2 Porosity

In order to be considered as a payrock, a rock body must be porous. Porosity is of paramount importance, for it determines the storage capacity of the reservoir rock.

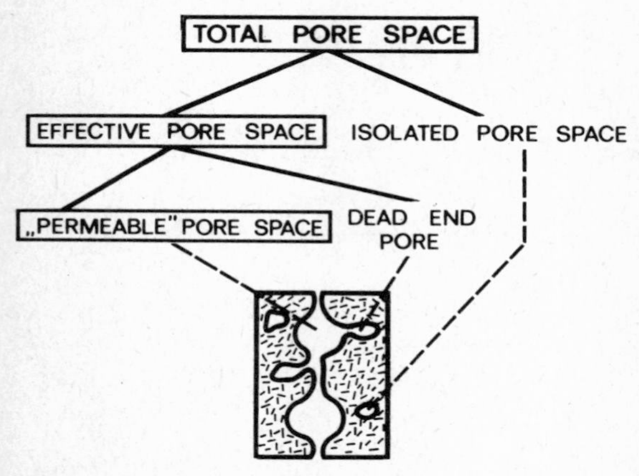

Fig. 8 Classification of Pore Space.
The total pore space is divided into effective and isolated pore space. The effective porosity is determined by interconnected void space. The isolated pore space is completely sealed off, having no pore channels connecting the individual cavities within the rock matrix. Only the portion of the pore space permitting flow of fluids contributes to conductivity. Dead end pores do not form pore channels, but play their part as pay volume. One has to be aware that any classification of the pore space is no more than an attempt to introduce an artificial order into the inconceivable complexity of the pore geometry

2.2.2.1 Classification

In oil-field practice, pore space is generally classified as represented in Fig. 8.

A good insight into the nature of porosity is given by a model using spheres. These can be either

a) balls of equal size (see Fig. 9), or

b) balls of differing sizes, with which the so-called grain sorting, that is to say size distribution, is decisive as far as porosity is concerned (Fig. 10).

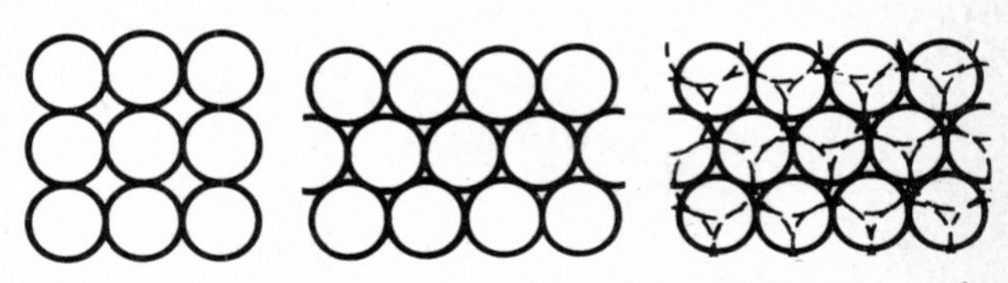

Fig. 9 Three possible ways of packing round balls of equal size. Theoretical porosities are between 25.9 % and 47.5 %

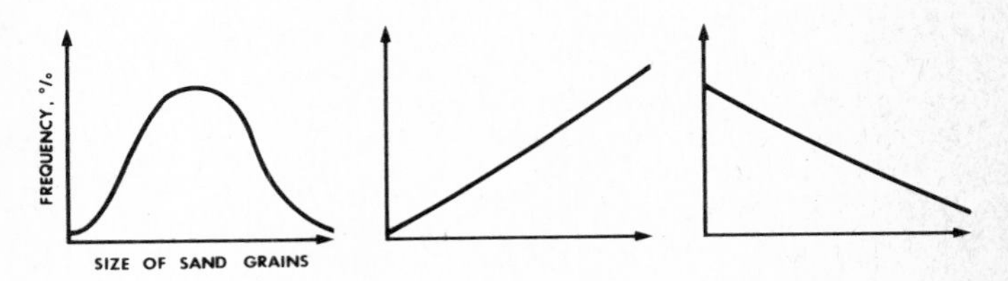

Fig. 10 a) Sands with a clear symmetric maximum of the grain size distribution show highest porosities. b) c) Sands with asymmetric grain size distribution show lower porosities

In reality, however, pay rock pore space is a very unevenly formed creation.

As far as the size of the cavities is concerned, we can distinguish between:

a) Normal pores which are so small that the fluids contained in the pores are subject to both gravity and capillary forces.

b) Sub-capillary cavities: The diameter of the pores is less than 0.002 mm and intermolecular forces work between the solid matrix and the pore content so that little flow occurs inside the pores.

c) Caverns: (Mostly found in igneous or karst rock), where fluid movement and cavity saturation is determined more by gravity than by capillary action.

2.2.2.2 *Definition*

The bulk volume V_b of a porous rock body is made up of its matrix volume V_m and its pore volume V_p:

$$V_b = V_m + V_p \tag{1}$$

The parameter which characterizes a porous medium is its porosity (ϕ), it informs us what portion of the bulk volume is filled with pieces of solid matrix material, and what portion remains as pore space:

$$\phi = \frac{\text{pore volume}}{\text{total volume}}$$

$$\phi = \frac{V_p}{V_b} = (1 - \frac{V_m}{V_b}) \tag{2}$$

The quotient of the two volumes is dimensionless. It is generally defined as "porosity factor f" or as a percentage of total volume (f = 0.25, ϕ = 25 %).

2.2.2.3 *Porosity of some rocks*

Porosity, as has been said, is dependent upon grain sorting and grain shape. Shales, for example, have framework of feldspar flakes which, as would be the case with playing-card houses, exhibit a high porosity. Sandstone, on the other hand, would be better compared with numerous balls fitted closely together. Limestone often has characteristics of both sand and shale.

Sand: sand porosity may be limited
by mechanical diagenesis
by chemical diagenesis
by pore compression, (due to increased overburden pressure resulting from the withdrawal of oil and gas).

The compressibility of sandstone is reversible: when a drilling core is lifted from great depth the pore space in the rock "increases".

Shales are to start off with very porous, but suffer a drastic loss of porosity due to compaction following the deposition of overlying rock layers. As opposed to compaction in sands, this compaction is irreversible. This fact can help in the reconstruction of the geological history of an area when, for example, a compacted shale is once again uplifted to shallow depth after compaction has occurred. The change of porosity and of permeability in context with the squeezing out of sedimentary water in shales can result in considerable drilling difficulties (overpressurized zones).

Limestones are especially susceptible to changes in porosity through chemical diagenesis. Here one must also account for fissures when measuring porosity.

Rock porosity varies within a wide range. In oil and natural gas payrocks, normal porosities range from 0–40 %. The average is 15–25 %. One way of determining average porosity is described in Appendix A.

2.2.2.4 *Measurement of Porosity*

The methods of porosity determination are now generally standardized and can be looked up in any textbook, so that only a number of general remarks will be made and some special cases discussed. A compilation is given in Appendix B.

a) From rock samples

In order to calculate the porosity of a rock body, two of the volumes from equation (1) must be measured in the lab; the third can then be calculated.

The bulk volume V_b can be ascertained through the use of geometrically simple bodies, with smooth surfaces (for example: a square or cylinder). These can still be simply measured with a measuring stick, but normally the total volume V_b must be determined by fluid saturation. In order to test for matrix and pore volume, one has to use liquids and gases, due to the complex labyrinth system of pore spaces within the rock. These mediums are, of course, the only ones which can accommodate to the contorted shapes of the pore passages to be measured.

Fig. 11 Strongly enlarged lead cast of a pore of a sandstone. (After *Collins*[5])

Using liquids:

To test for the total matrix and pore volumes of a rock the liquid volume displacement must first be identified

a) if the body under study is submerged in liquid without allowing the liquid to penetrate one can determine V_b (the body is first covered in mercury).

Fig. 12 A Valanginian Sandstone of a German oil-field with a porosity of 19 % as seen under the electron scanning microscope. Enlargement 344 × (upper) and 710 × (lower). (After *Gaida et al.*[10])

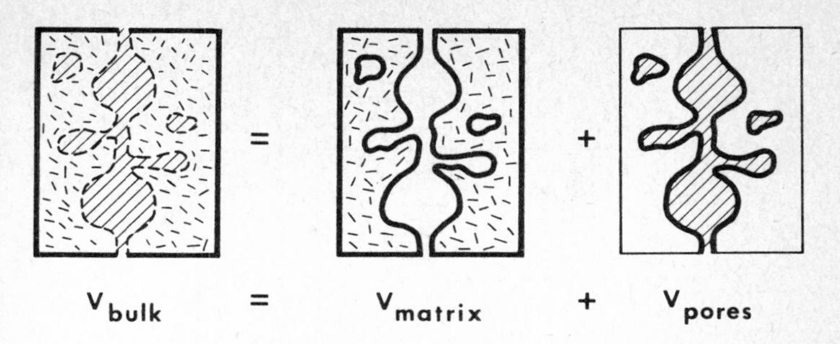

Fig. 13 The bulk volume V_b of a porous rock body is made up of its matrix volume V_m and its pore volume V_p

$$V_b = V_m + V_p$$

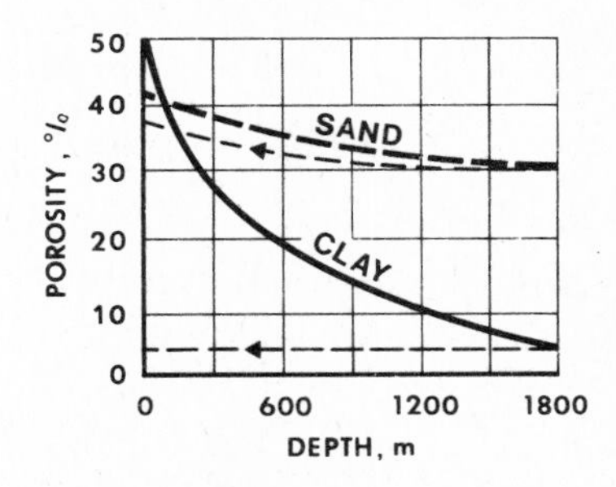

Fig. 14 Decrease by compaction of the porosity with increasing depth (after *Frick*[8]): The compaction of sandstone is reversible, that of clay is irreversible. In other words: the porosity of clay in this figure decreased to 4 % at a depth of 1800 m. A core drilled at this depth and brought to the surface shows the same porosity of 4 % as it is at a depth of 1800 m

b) If the pore volume can be filled with liquid (if the body can be "saturated"; the pore spaces must be completely filled), then V_m can be measured.
Using Gas:
In a volume-measuring instrument the body in question is placed in a closed chamber; this chamber is then filled with gas (Gas-pyknometer). From the change in condition (pressure or volume) of the gas after the emplacement of the sample, one can measure the matrix volume.
A simple empirical method of porosity determination on a thin section is described in Appendix C.
b) From well logs

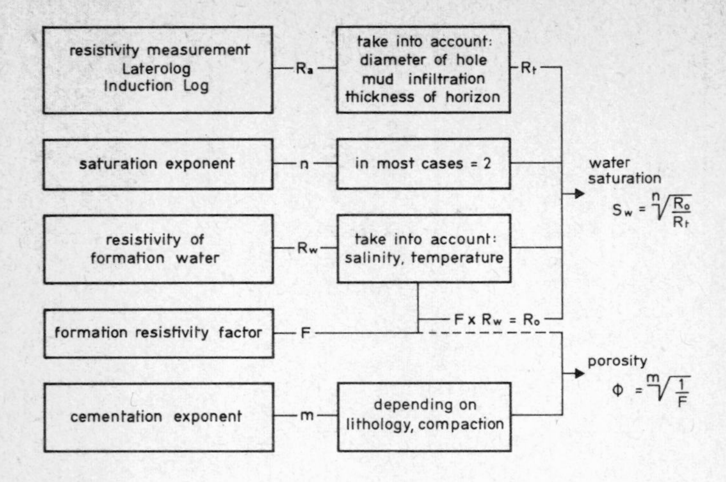

Fig. 15 Schematic representation for calculation of porosity ϕ and of water saturation S_w from well logs

Well logs are an indispensable tool not only for stratigraphical correlations and for the solution of geological and petrophysical rock problems. They are also a reliable tool for the determination of porosity and the saturation of reservoir rocks.

Porosity can be determined by the following logs:

1. Microlog (ML/MD)
2. Microlaterolog (MLL)
3. Neutron-Log (N)
4. Sidewall-Neutron-Porosity Tool (SNP)
5. Compensated Neutron Log Tool (CNL)
6. Sonic Log
7. Formation Density Tool (FDC)

2.2.2.5 Pore Compressibility (Soil Subsidence)

Immediately following deposition, the weight of overlying sediments causes compaction. This results in a reduction of the total or bulk volume V_b as well as a reduction in pore volume V_p. This volume reduction is especially large in shale (see Fig. 16) and lime muds; it is, however, smaller in sands and oolitic limes. In effect, the volume reduction in pore space involved in this process is a

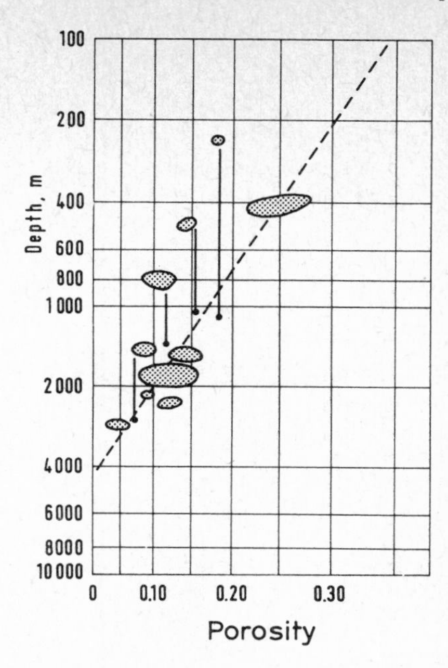

Fig. 16 Porosity in various oil-fields (dotted areas) of a German sedimentary basin. As compaction of shale occurs, a logarithmic progression is produced between depth of burial and porosity. This dependency is shown by the dotted line. With the help of this depth-porosity line one can often reconstruct an area's geological history, because the compaction of shales is irreversible (see Fig. 14). For example, in the above figure Liassic shale from a depth of 400 m has a "normal porosity" of 25 %; the same shale at 3000 m has an equally "normal" porosity of 5 %. The porosity of a few fields lies distinctly outside, e.g. above this line. Their porosity is larger than the "normal" value. This shows that the structural position of the Liassic shale has changed through later tectonic movements. In the above case through salt movement (Halokinese) the shale was structurally lifted. (After *v. Engelhardt*[6])

reduction in porosity. In a similar manner, rock compaction also influences permeability (see Fig. 17).
Of special importance is the compressibility of the pore space. This occurs with the production of oil and gas when pressure within the pores decreases. The pore compression caused by the withdrawal of oil and gas may lead to considerable soil subsidence and surface damage (see Fig. 18 and 19).

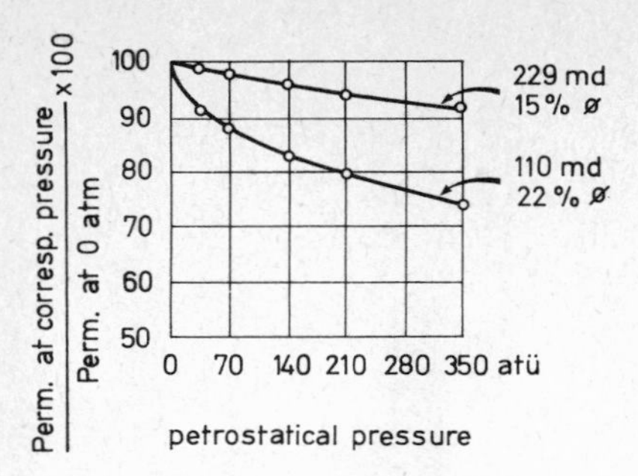

Fig. 17 The pressure of overlying rock layers (petrostatic pressure) and the resulting rock-compaction brings about a reduction of the porosity as well as the permeability (after *Fatt*[7]). The extent of this reduction depends upon rock type

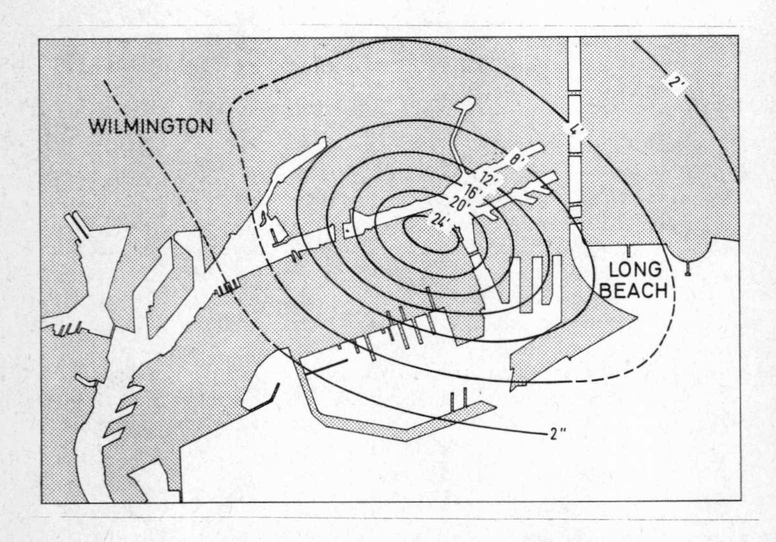

Fig. 18 The pore compressibility can be so great that it leads to extensive amounts of soil subsidence over an oil or gas field. A well-known example is the Wilmington field in California where an area of 40 km² (the total field area) was affected by soil subsidence. At its deepest, the ground had sunk a full 8 m, so that considerable damage occurred to houses, bridges etc.

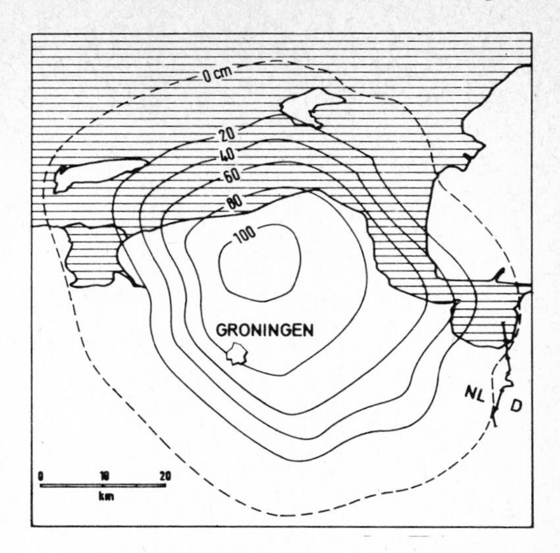

Fig. 19 In a theoretical calculation it has been figured for the Dutch gas field Groningen that as much as 100 cm subsidence can be expected at the field centre[23]

For this reason, detailed examinations and calculations have to be made[13]: the total or rock compressibility c_b consists of

c_s = matrix compressibility of sandstone
$+ c_p$ = pore compressibility
$= c_b$ (rock compressibility)

Matrix compressibility is a linear function of pressure and is somewhat larger in sandstones (2.65×10^{-6} cm²/kg) than in limestones (1.37×10^{-6} cm²/kg) (see Fig. 20, 21 and 22).

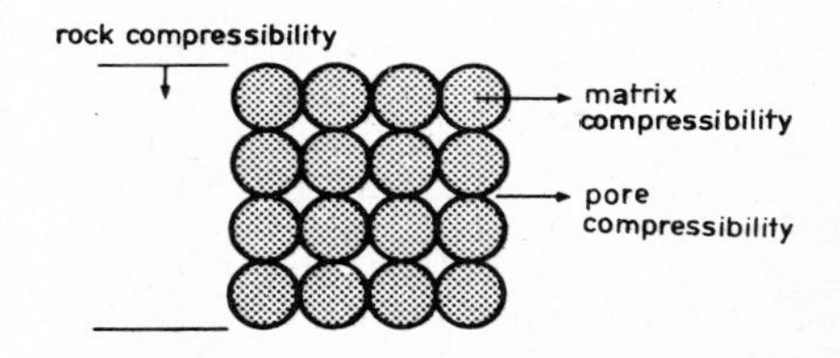

Fig. 20 The total compressibility of a porous rock is made up of matrix compressibility and pore compressibility

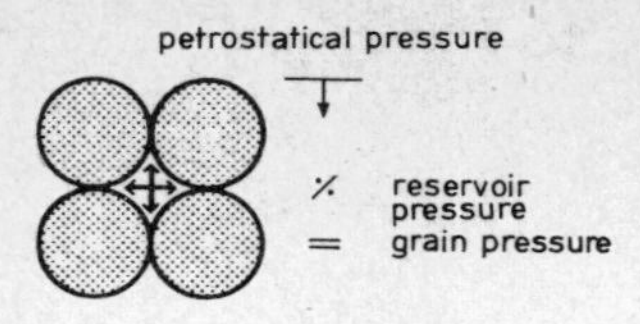

Fig. 21 The reservoir pressure prevailing in the pores is "counteracting" the petrostatic pressure from above: The difference between the petrostatic and reservoir pressure is called grain pressure. With the beginning of oil or gas production the reservoir pressure begins to decrease, but pore space compression rises and therefore the grain pressure rises. Compression of the pore space then occurs

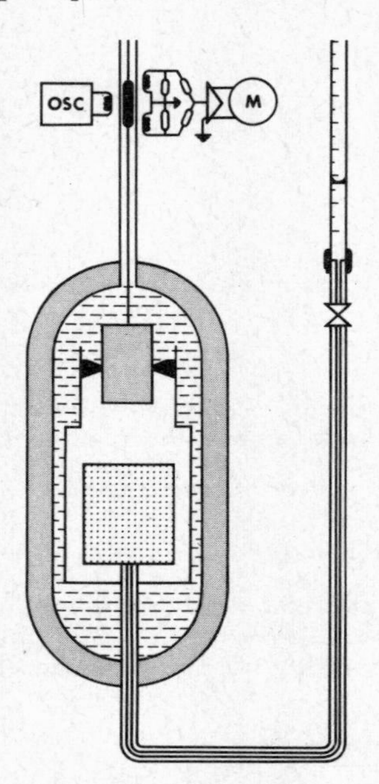

Fig. 22 The essential factor is the so-called "effective pore compressibility c_f"; and this depends on "outside" petrostatic pressure, and "inner" pore liquid pressure (or reservoir pressure). Inner pressure of course changes with the withdrawal of oil and gas from the reservoir rock. The difference between "inner" and "outer" pressure is called "grain pressure". Instruments have been developed to examine this relationship. And it is through this examination that the "effective pore compressibility" for any desired reservoir pressure can be determined. (After *v. d. Knaap*[13])

2.2.3 Permeability

A reservoir rock must be capable of storing hydrocarbons (oil and gas) and of transporting them through itself; these qualities of the rock are porosity and permeability. Porosity gives a measure of the volume stored in the reservoir rock. Permeability is the yardstick for the flow of fluids through the rock. A rough and very schematic comparison: the rooms of a house can be compared to the pore space, the connecting doors and corridors can be compared to the permeability. Both are overall quantities and permeability + porosity might give a better insight into the structure of the reservoir rocks.

2.2.3.1 Darcy's experiment

This experiment (in connection with water supply to the city of Dijon) by the Frenchman Darcy (1803–1858), who used an artificial sand body and a constant water volume, was a simple case: homogeneous, linear, isopermeable, steady-state, incompressible (see table p. 79). The experiment showed that the amount of water flowing through in a given time depends on length L of the sand body, water level at inlet and outlet and a constant K.

$$q = -K \frac{h_1 - h_2}{L}$$

Later on, the oil industry, recalling these experiments, repeated them with greater precision and, after finding them confirmed in all respects, put them into the following formula:

$$q = \frac{k\,A\,\Delta p}{\mu L},$$

where: L = length of rock sample
A = cross-section of rock sample
Δp = pressure differential
μ = viscosity of flowing medium
q = flow per unit of time
k = permeability constant.

(In practical lab tests, the density of the flowing medium ϱ and the gravitational constant g may be neglected.)

2.2.3.2 Definition

Now this rock factor, the "permeability constant k", had to be defined by giving it a numerical value. The definition is wholly empirical, designed to meet practical requirements, and reads as follows: if a liquid having a viscosity of 1 cp (μ) flows through a porous rock of 1 cm length (L) and 1 cm^2 cross-section (A) at a

rate of 1 cm³/sec (q) when the pressure differential between inlet and outlet is 1 bar (Δp), then, according to the definition, the rock's permeability (k) is 1 Darcy.
The above-mentioned units, established by consent, therefore give the permeability in Darcy ($\frac{1}{1000}$ Darcy = 1 millidarcy = 1 md).
Besides this, the so-called "Perm system" is still used in some American oil-field operations, although it has not been generally introduced.

	Darcy system	cgs	Perm system
L	1 cm	cm	ft
A	1 cm²	cm²	ft²
Δp	1 at	dyn/cm²	psi
μ	1 cp	p	cp
q	1 cm³/sec	cm³/sec	1 bbl/d

$$1 \text{ darcy} = 10^{-8} \text{ cm}^2 = 0.88 \text{ Perm}$$
$$1 \text{ Perm} = 1.13 \text{ darcy}$$

When analysing the dimensions, it appears that the permeability k in the cgs system has the same dimension as an area (length²).
Darcy's law is valid only if some physical conditions are met:

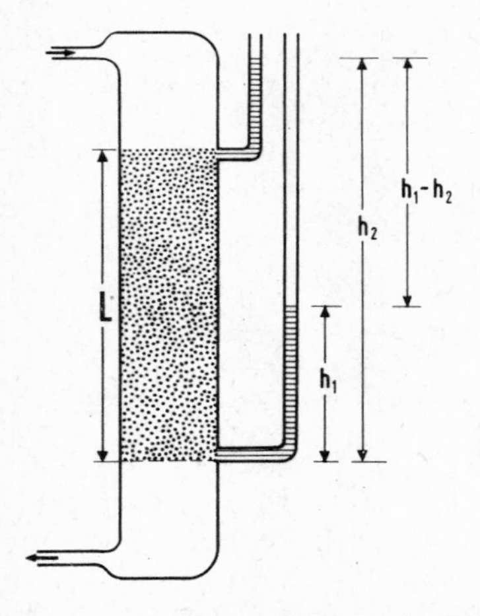

Fig. 23 Darcy's experiment (after *Hubbert*[11]) showed that the amount of water flowing through an artificial sand body in a given time depends on length L of the sand body, water level at inlet (h_2) and outlet (h_1) and a constant K

a) Reynold's number
The application of Darcy's law is restricted to laminar flow, where the flow behaviour is only governed by the forces of (internal) friction (viscosity), while inertial forces (causing turbulence phenomena) do not yet play a part.
b) Knudsen flow and Klinkenberg effect
Permeability is a matrix-specific, i.e. rock-related, material constant, which no longer depends on the state of the pore content. It should therefore be possible to describe the flow of liquids and gases by a uniform permeability. This applies to the flow behaviour of liquids which are inert with respect to the matrix of the porous medium; likewise for gases under high pressure (see also page 81).
c) Special conditions occur when electrolyte solutions flow through a rock which contains clays, in particular montmorillonites resulting in deviations from Darcy's equation.

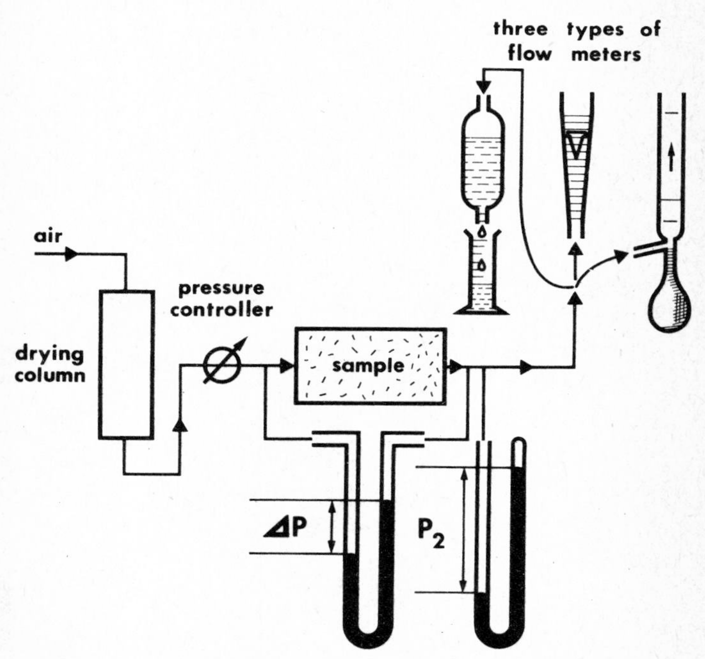

Fig. 24 Schematic diagram to measure the permeability on a rock sample

2.2.3.3 *Measurement of Permeability*

The experimental determination of permeability is mostly done on cylindrical samples. If anisotropy effects are to be investigated in

particular, it is advisable to measure the permeability on cuboidal samples in the direction of the three axes of the cuboid. Before measurement, the dried sample should be mounted in a suitable holder which prevents bypasses for the gas or liquid stream along the sample surface (Fig. 24).
Permeability is preferably determined during steady-state flow. There are also methods, however, which enable the permeability to be calculated from unsteady-state flow.

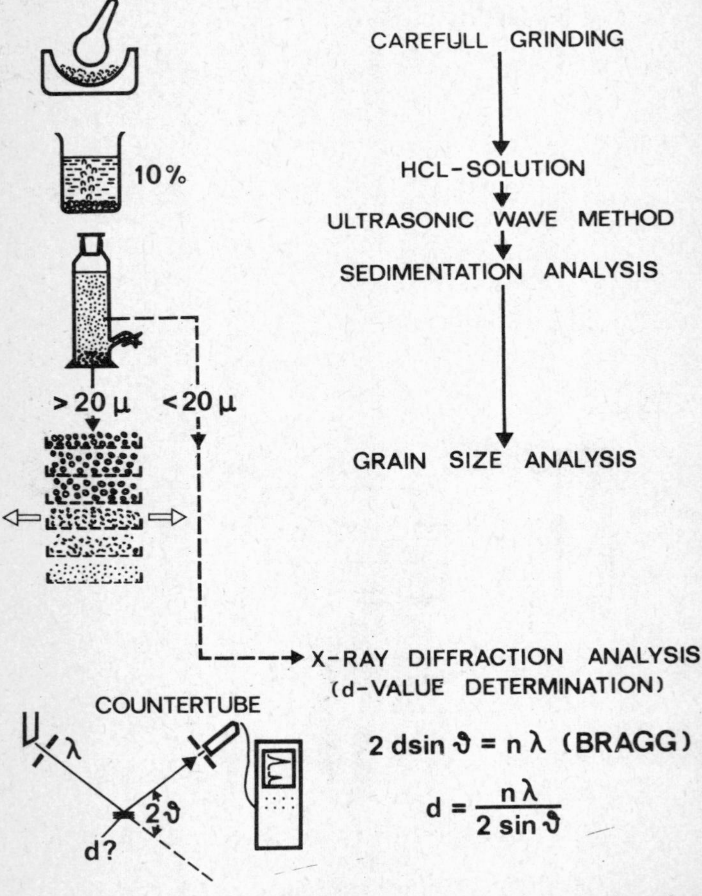

Fig. 25 Schematic representation of the rock matrix investigation of a sandstone (by courtesy of Dr. *Miessner*, Brigitta/Elwerath, Hannover)

Moreover, it is possible to compute the permeability in the drainage area of a well from measurements of pressure decline and build-up in the borehole (see page 127). These measurements are also based on unsteady-state, although in this case radial, flow conditions. From such pressure build-up and decline processes it is also possible to calculate the permeability capacity (k · h) of rock formations. "Guestimating" of permeability from P_c curves see page 42.

2.2.4 Other Parameters

Besides the best known and most important physical payrock parameters, that is porosity, permeability and capillary pressure, there are also a number of other factors which are of some interest to the petroleum engineer. These have to do, in part, with theoretical considerations.
In Fig. 25 the preparation and examination of sandstone matrix is described. (The HCL treatment should be used only along specific guidelines and only when an acid treatment is suitable.) Ultrasonic treatment should be regarded as a sort of shock treatment used to separate clay particles from sand grains.
a) Grain Size and Shape
These parameters play an important part as far as porosity is concerned. They may be ascertained
either through *sieve-analysis,* in which case grain shape is only indirectly determined through grain size (through the use of a "shape factor" which helps determine deviation from a completely round form);
or by the *sedimentation procedure,* which is based on the fact that radius is a clearly representative parameter for the shape and size of a sphere. Large spheres sink more quickly than small ones. Furthermore, the more jagged and uneven the grain is, the slower it falls (and therefore the smaller is its breadth in relation to a perfect sphere).
A graphic representation of the grain size frequency distribution helps to shed light on many of the petrophysicist's problems (see Fig. 10 on page 9).
b) X-Ray Diffraction Analysis
The X-Ray Diffraction Analysis is a useful tool and is popular with petrophysicists since it helps to identify crystalline substances. For details see appendix D.
c) Pore Radius
For the judging of pore space, this can serve as a valuable tool; this subject is discussed along with capillary pressure measurements on page 42.
d) Inner Surface Area
The "inner surface area" $\bar{S}$ is given in cm^2 (or mm^2). The "specific

inner surface area" S is a factor in the relationship to the pore volume:

$$S = \frac{\bar{S}}{V_p}$$

In comparison with such absorbant materials as charcoal or silica gell, the inner surface area of pay rocks is small. It is, however, an important parameter, especially when estimating connate-water content. The parameter's value may be ascertained through the use of either a thin section (approximately the same procedure as described in appendix C for porosity determination) or a grain size distribution (only when dealing with similarly shaped grains where the shape factor is known).
The most widely applied method, however, is the BET-method. This method notes the amount of N_2 or CO_2 molecules absorbed and then using the gas laws calculates $\bar{S}$. One can also use the Kozeny-equation (page 27) to determine the inner surface area.
e) Tortuosity
The tortuosity T is an expression for the relationship of a rock body L to the length of the capillary cavity l_{eff} in the rock body. This capillary cavity is, however, not a smooth pipe but is rather a intricately built passage. Therefore one can also use the more descriptive expression "detour factor" (Fig. 26).

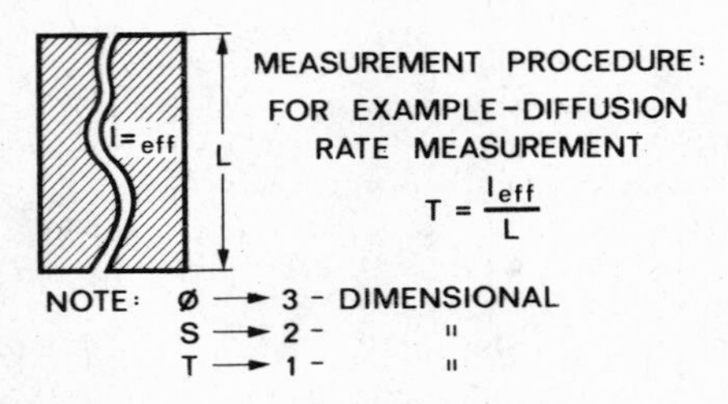

Fig. 26 Tortuousity (Detour Factor) T

As the relationship $\frac{L}{l_{eff}}$ is also reflected in the Kozeny equation, (see page 27) one may also observe the relationship between porosity and tortuosity in this equation.
A similar relationship between T and ϕ is also included in the formation resistivity factor F, (relationship between the resistivity of a salt water saturated rock sample R_o and salt water itself R_w. (See page 28). This formation resistivity factor F can now be seen in term of Archie's formula which describes the relationship of F to porosity as:

$$F = \frac{1}{\phi^2}$$

which in terms of tortuosity becomes

$$T = \phi^2 \times F^2$$

2.2.5 Relationship between the Individual Parameters

In order to get a complete view of the extremely complicated geometry of the pore space, one could make the pore space visible through the use of a thin section or a plastic impression (see Figures 11 and 12). There are also other ways to make the pore space accessible for description. But in spite of much progress that has been made in recognizing the geometry of the pore space, many individual questions are still open. To use a simple analogy: If the building one is at present sitting in were compared to a reservoir rock, then one could easily determine how much usable space the building contained by estimating how big the rooms and passages were, on the average. But one would still not know the exact dimensions of the individual rooms and how each room looked.
Of course what would obviously be expected would be a relationship between permeability and porosity; a mathematical or statistical relationship has been extensively sought after. A permeable rock body always exhibits a certain porosity, but on the other hand

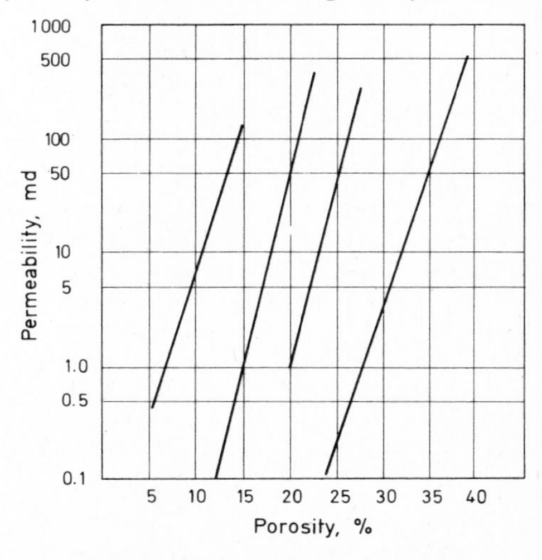

Fig. 27 Porosity in relationship to Permeability. Graphic representations usually show porosity and log-permeability in terms of their relationship to a specific rock type

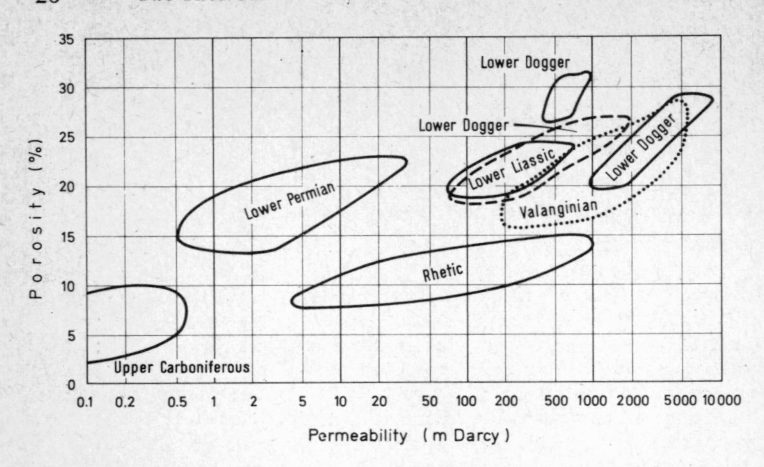

Fig. 28 Relationship between porosity and permeability of NW-German sandstones of different geological age. (After *Gaida* et al.[10])

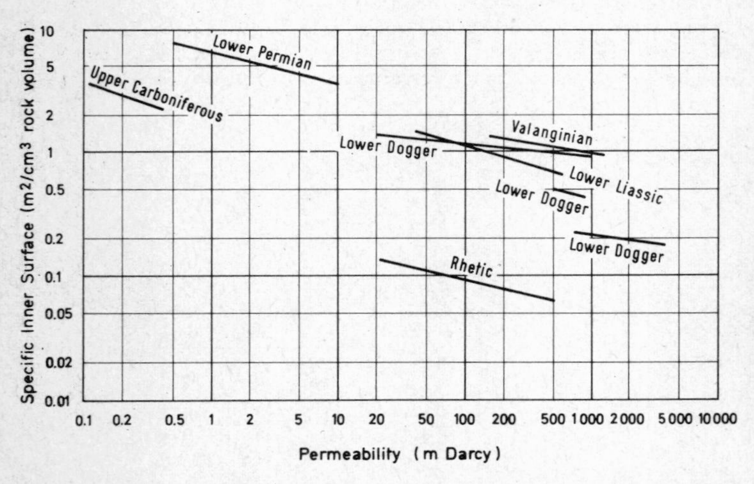

Fig. 29 Relationship between Permeability and Specific Inner Surface of NW-German sandstones of different age. (After *Gaida* et al.[10])

there are many rock bodies which, although porous, have comparatively low permeability. Or, to use another example, a coarse grained sandstone and a piece of writing chalk have the same porosity. But the writing chalk is almost impermeable (k = 1 md),

while the sandstone may show permeability values higher than 10,000 md. Porous but non-permeable gas bearing rocks are a suitable target for atomic subsurface blows.
A similar relationship exists between permeability and the inner surface of a payrock. See Fig. 29.
Clearly in contrast is the relationship between permeability and S_w; the smaller the permeability of a rock, the greater its connate-water content is (Fig. 30).

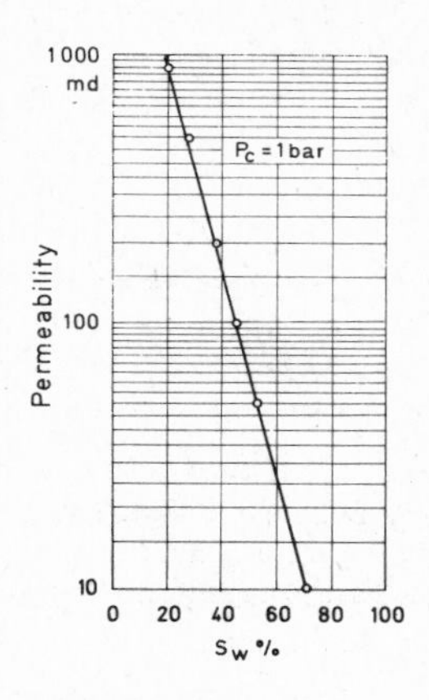

Fig. 30 Permeability and Water Saturation. The illustration shows a linear relationship in which log-permeability and S_w are plotted against each other

The Kozeny-Carman Equation is an attempt to obtain an understanding of pore space using mathematical methods. Let us conceive of a model simple enough to be thought of mathematically and general enough to include all possible pore cavity shapes. With the help of the hydraulic radius m, which describes the relationship between inner surface area and interior volume in the capillary cavities, one brings into focus the relationship between the Darcy equation, porosity and inner surface area[16].

2.2.6 Electrical Resistivity of Sandstones

Electrical resistivity measurements of partly or fully water-saturated sandstones have, for decades, been an integral part of well logging. For that reason interest has, for a long time, been shown in all parameters describing electrical conductivity in fluid-saturated porous rocks. This of course helps to interpret well logs, so that porosity and water saturation can be accurately determined (see page 14).

a) Formation Resistivity Factor "F"

In Fig. 33 a current (I) flows through a completely water-saturated rock sample. As the rock matrix is not conductive, the current must

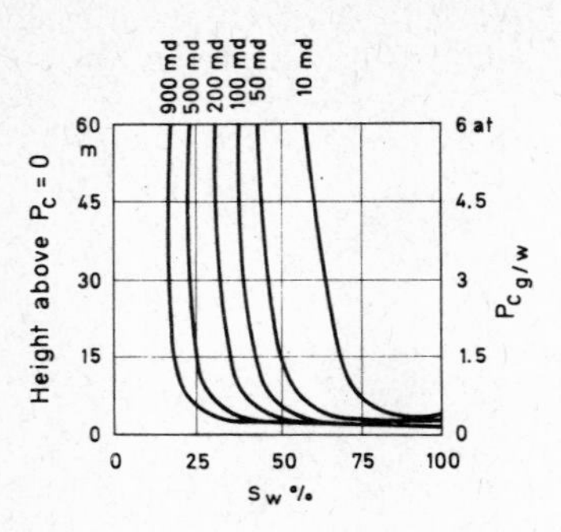

Fig. 31 Permeability and S_w. The illustration (after *Wright* et al.[24]) demonstrates that a poorly permeable rock body with a permeability of 10 md shows, at a height of 45 m, a saturation S_w of 60 %. While a highly permeable rock body (900 md) at the same height has only a 15 % S_w

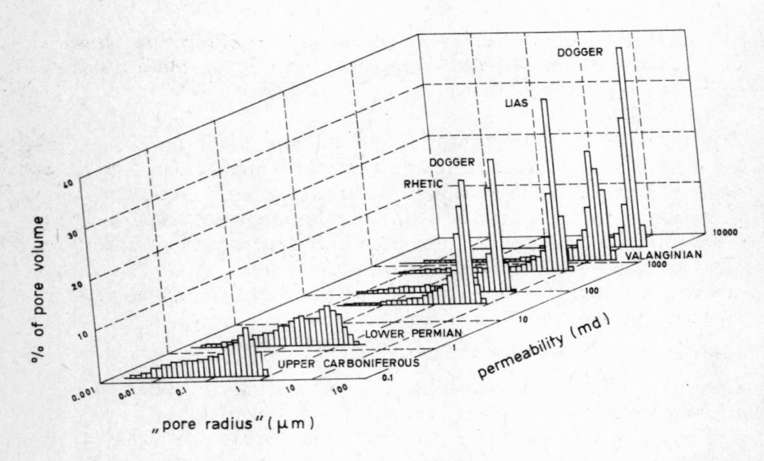

Fig. 32 Relationship between porosity, permeability and "pore radius" demonstrated on reservoir rocks of different geological ages. (After *Gaida* et al. [10])

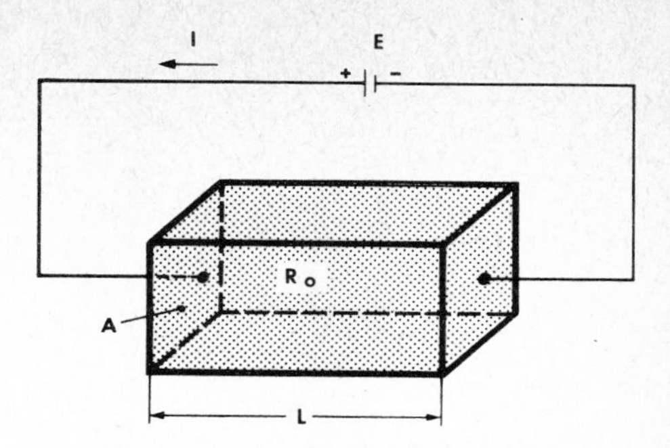

Fig. 33 If an electrical voltage (E) is attached to opposite faces of a rectangular rock sample which is completely saturated with salt water the current (I) flows through the sample. The resistivity of the water saturated rock is $R_0 = \frac{E}{I}$. (After *Pirson*[17])

flow through the salt water – the "electrolyte" – that fills the pore spaces. According to Ohm's law, the electrical rock resistivity is:

$$R_0 = \frac{E}{I}$$

This resistivity is not a "rock-constant" but rather depends on the specific resistivity of the salt water and therefore on the ion-concentration of the electrolyte (salt water). (With a high ion-concentration, the resistivity is low; if the ion-concentration is low, resistivity will be high.)

A container with the same dimensions as the rock sample (length L, cross-sectional area A in fig. 34) filled with salt water (of the same salinity as the water saturating the rock sample) shows a resistivity R which can also be calculated using the specific resistivity (R_w) of the salt water:

$$R = R_w \cdot \frac{L}{A}$$

The ratio $\frac{R_0}{R}$ represents a rock-constant parameter: the Formation Resistivity Factor F.

$$F = \frac{R_0}{R} = \frac{R_0}{R_w \cdot \frac{L}{A}} = \frac{R_w}{R_0} \cdot \frac{A}{L}$$

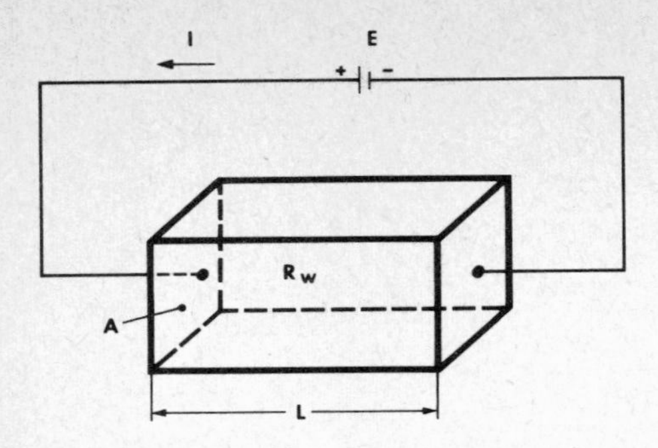

Fig. 34 The electrical resistivity of a container filled with water (of a specific resistivity R_w) is $R = R_w \cdot \frac{L}{A}$. (After *Pirson*[17])

Apart from R_o and R_w this formation resistivity factor depends on the pore space geometry,in other words: on the porosity, the cementation exponent m and the packing of the sand grains (see Fig. 35). To determine porosity the best known (empirically found) formulae are

carbonates $\frac{1}{\phi^2}$, Archie Formula $\frac{a}{\phi^m}$ $= F =$ $\frac{0,62}{\phi^{2,15}}$ Humble Formula (sands, sandstones), $\frac{0,81}{\phi^2}$ sands

Archie's Formula $F = \phi^{-m}$
Humble Formula $F = 0.62\ \phi^{-2.15}$
$m = 1.3 \ldots 2.2$

b) Resistivity Index I

If the pore space is not completely filled with electrolyte, but partly with a non-conductive medium (such as oil), the electrical resistivity will be higher than R_o. This relationship between the two resistivities is called the "Resistivity Index I".

$$I = \frac{R_t}{R_o}$$

The resistivity index is determined by the saturation in the pore space of the electrolyte (water) and the non-electrolyte (oil). So the wettability of the fluids is an important factor.

When interpreting the well logs the empirically found correlation is used:

$$I = \left(\frac{1}{S_w}\right)^n$$

(n, the saturation exponent, is generally between 1.8 and 2.0).

c) Shaly Sandstones

Many sandstones contain clay minerals. These shales may occur

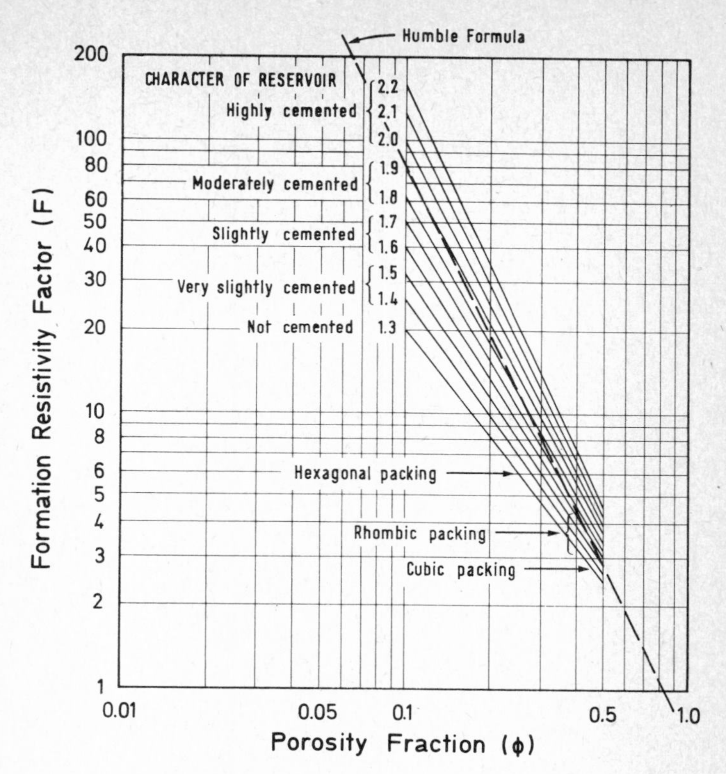

Fig. 35 The Formation Resistivity Factor F is the expression of a property of reservoir rocks which is an indispensable parameter for quantitative well log interpretation. F is dependent on porosity, cementation and the packing of the sand grains. (After *Pirson*[17])

either as layers, or as shale lenses in the sandstone or even as clay minerals in the pore spaces of sandstones. Since clay minerals are active electrolytes, the electrical properties of shaly sandstones are highly complex. Obviously the electrolytic activity of the clay minerals raises the conductivity of the rock (although this increase itself is a function of the fluid salinity).

Sandstones having a higher clay content (laminar or diffuse) therefore falsely indicate a greater water saturation of the pay horizon. Likewise all porosity measurements in the borehole are influenced by increasing clay contents, in that they show too high porosity values. The "Waxman-Smits-Correlation" is a good quantitative description of the electrolytic conditions in shaly sandstones and definitely improves the interpretation of well logs.

2.3 The Content of the Reservoir

2.3.1 Saturation

The pore content is generally called "saturation" S; the substances in question being gas, oil and water as symbolized by S_g, S_0 and S_w respectively. It is very important to know the saturation of a reservoir rock. This goes not only for the calculation of reserves but also for non-quantitative analysis in connection with e.g. flow, production, planning, well logging etc.
The saturation is determined either
from drilling cores or
through geophysical well logs or
with the help of capillary pressure curves and
various other methods.

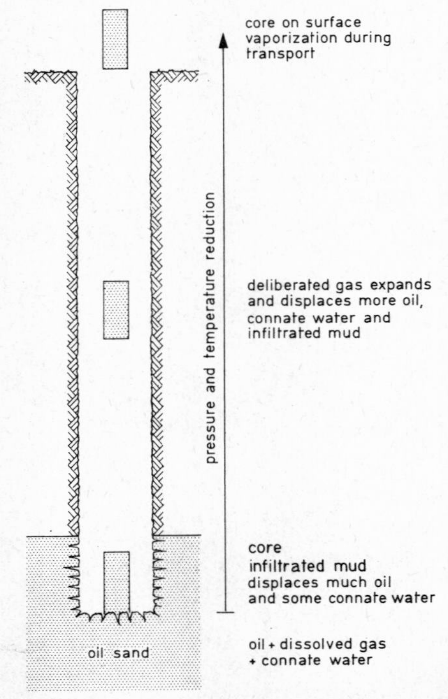

Fig. 36 What happens to an oilsand core on its way from the bottom of the hole to the lab?

2.3.1.1 Measurement on Core Samples

Normally the determination of a core's pore content causes, in principle, no difficulties. But, as far as the drilling core is concerned, it is important to consider that between the time the core is drilled and the time it is received at the lab, substantial changes occur. These effects and their consequences are shown in Fig. 36. Obviously the sample no longer represents the real subsurface reservoir conditions.

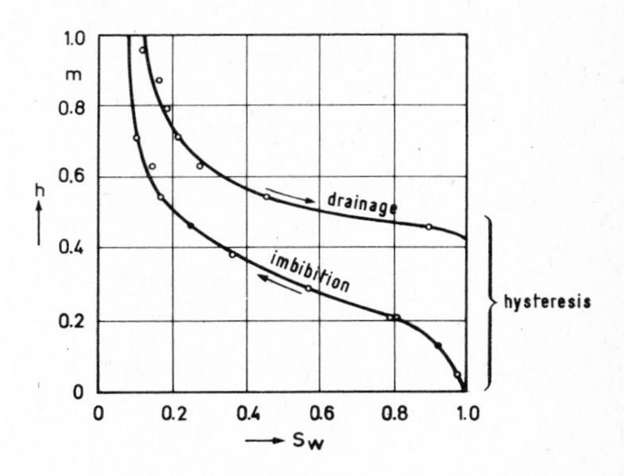

Fig. 37 Water saturation S_w at different levels h (in meters) of a core sample from a German Valanginian sandstone (after *Becker*[3]). Saturation values obtained by the drainage and imbibition methods. Theoretically the two methods should give the same results. The drainage method gives higher S_w values than the imbibition process. This difference, at times considerable, is generally called the hysteresis effect

Imbibition and Drainage:
If one were to dip a long, dry drilling core from a reservoir rock into water, capillary forces would begin to build up in the pore space, and after a longer time-span one would achieve an equilibrium between capillary forces and gravity. With the help of a special device which permits the measurement of the electrical resistivity of the rock sample one can calculate the water saturation at different heights. Therefore it is necessary to use NaCl-saturated water. Another way is to saturate the core with NaCl water initially, and then allow excess water to drip away until an equilibrium between gravity and capillary forces is once again attained. (With the formula $P_c = h\,(\varrho_w - \varrho_a)$ one can figure out the capillary pressure P_c. The Hg injection method serves as a good control and double-check).

Other Methods:
When the water-saturated core is centrifuged, the residual water can also be measured.
For the diaphragm method see Fig. 38.

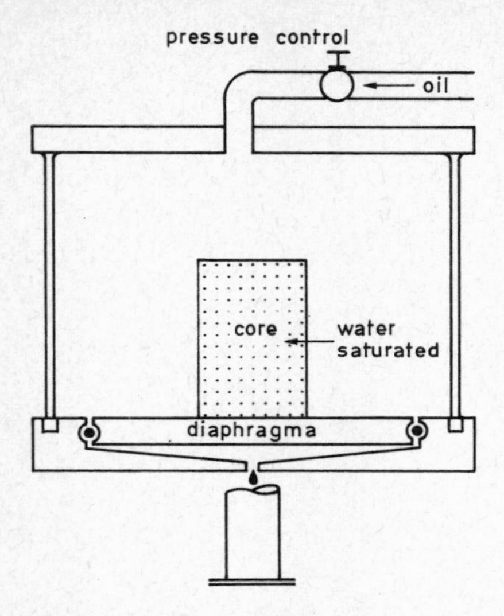

Fig. 38 Measurement of Saturation.
With the diaphragm method the drilling core is at first fully water-saturated, then the water is displaced by oil. The drained water may then be measured, as the porous, semi-permeable diaphragm lets through only water but no oil

The calculation of S_w from well logs is schematically shown in Fig. 15.

2.3.1.2 Capillary Pressure P_c as a Help to determining Saturation

An understanding of the saturation relationships is only possible after considering the fact that the pores are cavities of capillary size. Therefore capillary pressure is, besides gravity, an effective force in the pores. Together with wettability, these forces affect the often surprising distribution of gas, oil and water in the reservoir rock pores throughout the different parts of a geological structure.

Wettability:
When a fluid or gas comes into contact with a solid object, cohesion (the attractive force between individual molecules) and adhesion

(attractive force between molecules of different material) act together. When a gas or a fluid meets a solid body, it either spreads out over the whole body, "wetting it"; or it forms a well-rounded drop and wets the body either very little or not at all.
It is not yet quite clear what is meant by "wettability condition" or "degree of wettability" in a reservoir rock. Wettability changes from field to field and is difficult to predict. Normally, when water and crude oil fill the pore space, the water is the wetting liquid. When oil and gas are present, then oil is the wetting medium. Attempts to change the wettability conditions (for example through temperature or chemical procedures) are being continually made and provide hope for useful and interesting results.

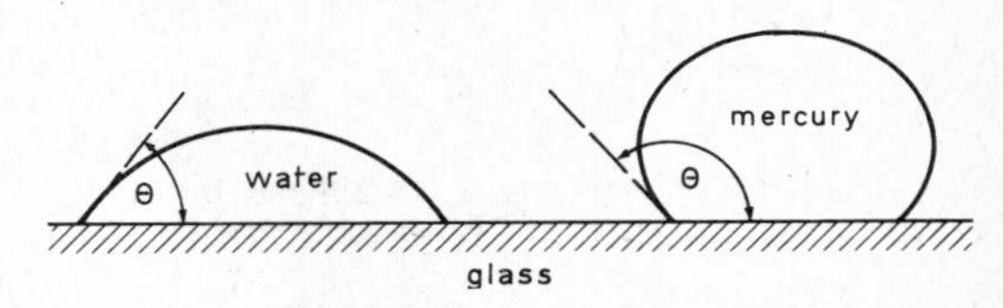

Fig. 39 If the angle of contact θ between the solid body and the fluid is smaller than 90°, then the fluid "wets" the contacted body. If the angle of contact is greater than 90°, the liquid is "non-wetting"

Capillary Pressure P_c:
Liquids, when in contact with gases, as a result of their internal molecular cohesion, keep their surfaces as small as possible; the best known example is the raindrop. On the curved liquid/gas surface the pressure on the inside (concave) of the curvature is larger than on the outside (convex). This pressure difference (Δp) is proportional to the surface tension (σ in dyn/cm) and inversely

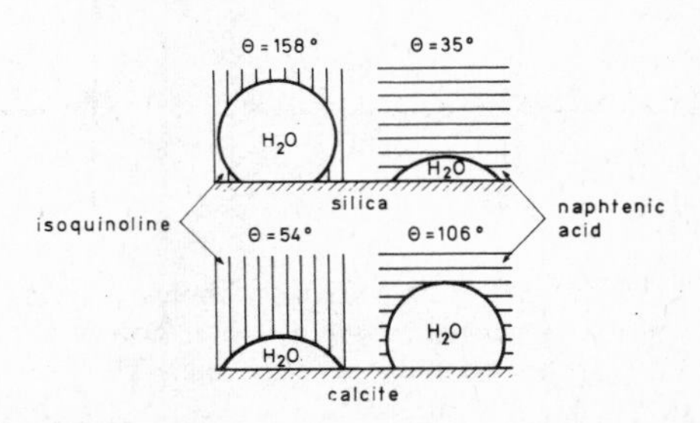

Fig. 40 The wettability is not only dependent upon the rock type (that is the mineral composition), but also upon the type of the liquid. (After *Amyx*[1])

proportional to the curvature's radius. This pressure difference is called capillary pressure (P_c), when the liquid filling the capillary cavities (radius r_t) is in contact with a solid body (see Fig. 41). Capillary pressure is known to raise wettable liquids in capillary tubes (see Fig. 42 and Appendix E).

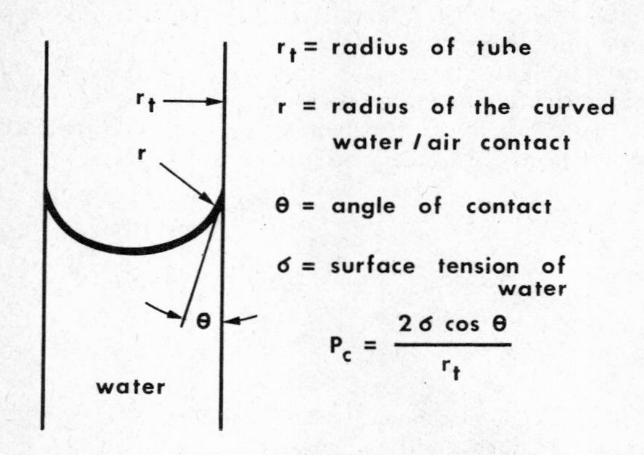

Fig. 41 The contact zone between water and gas in a capillary tube, is shown. The figure shows that capillary pressure is a clue to the pore width (r_t)

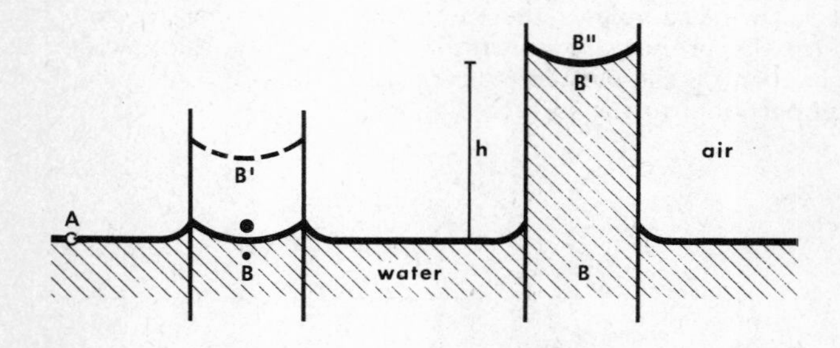

Fig. 42 Capillary pressure raises wetting liquids in capillary tubes. As we have seen, along a curved water/air interface, the pressure on the convex side is smaller than on the concave one. (The pressure difference Δp is equal to the capillary pressure P_c). Therefore within the capillary tube (at B on the convex side of the curved water/air contact) there is less pressure than at the liquid surface outside the tube (at A). The liquid rises by the height h from B to B′. The capillary pressure P_c equals $h\varrho_w$

Measurement of Capillary Pressure P_c:
To measure the capillary pressure of a porous rock, the mercury injection method is commonly used today, because it is the simplest and quickest method. It should be borne in mind that Hg is non-wetting with respect to the rock, so that it has to be injected under pressure into the reservoir rock. This injection pressure corresponds to the capillary pressure. (Fig. 43 and Appendix F.)

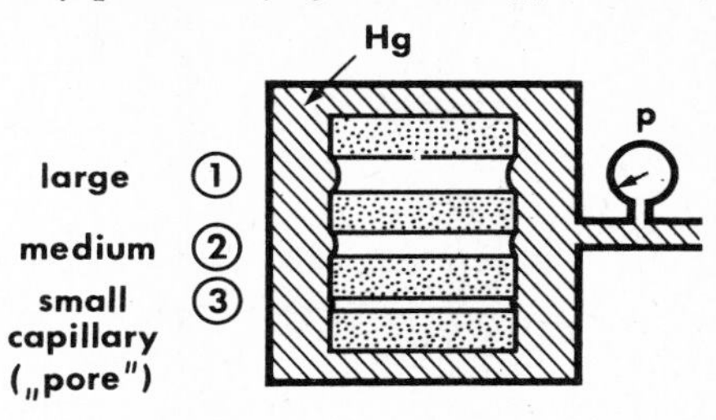

Fig. 43 Into a dried rock sample, which has been placed in a suitable pressure chamber, Hg is injected. During this process, at first the larger pores are filled with Hg, and then, with increasing pressure, also the smaller ones. The injection of Hg into a porous rock sample is represented in Appendix F, where also the conversion factors needed to change the Hg/air system into a water/oil system are given

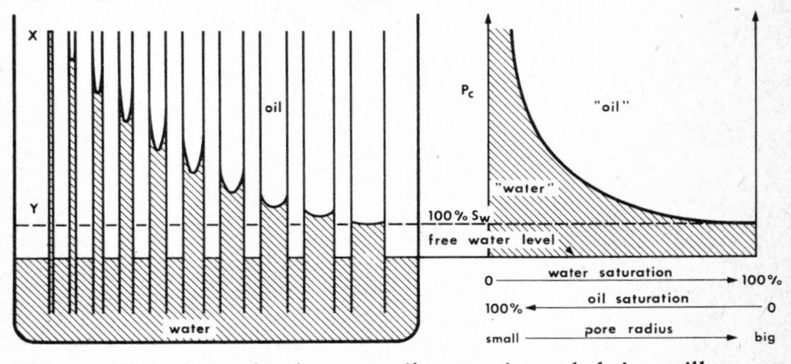

Fig. 44 The relationship between oil reservoirs and their capillary pore space can be schematically shown by a vessel containing oil above water and a bundle of capillary tubes ranging in diameter from very small to large. This pipe system would thus schematically demonstrate the pores of our reservoir rock. Now dip this bundle of capillary tubes into the water and note that the wetting fluid, that is the water, rises above the free water table. The smaller the radius of the tubes, the higher the water level rises and the smaller the curvation radii of the water/oil contact become. The capillary pressure P_c goes up accordingly in line with the formula on page 45

2.3.1.3 *Saturation and Capillary Pressure*

Figure 44 compares schematically an oil reservoir with a vessel containing oil and water and a bundle of capillary tubes of different diameters.

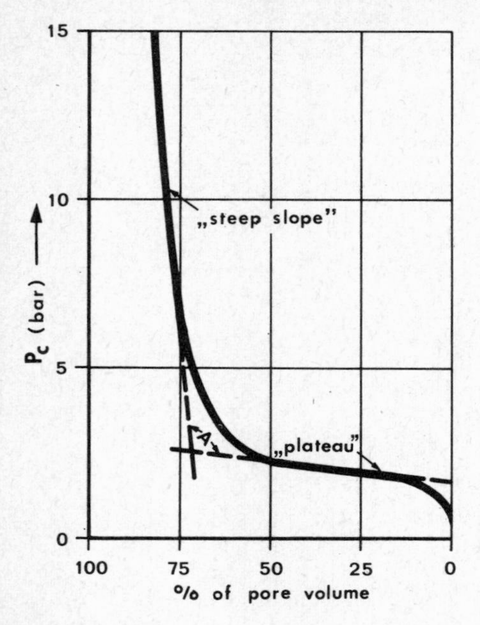

Fig. 45 An "ideal" P_c-Curve

In this illustration the capillary pressure curve of a well-sorted sandstone is shown. One sees that an initial pressure of 2 bar had to be applied until the first mercury penetrates into the rock body. This "displacement or threshold pressure" P_{cd} corresponds to the difference between the free water table and the line of 100 % water saturation shown in illustration 44. The smaller the permeability, or the poorer the sorting of a sandstone, the greater is this displacement pressure. Furthermore, one sees that the largest part of the pore space will be filled with mercury at a pressure between 2 and 2.5 bar. This section is called "the plateau". This plateau may always be clearly seen and lies relatively deep in sandstones with high permeabilities and good sorting. There follows the so-called "steep slope", where in spite of rising pressure, continually less pore space is filled with mercury; in the end, despite high pressure no further invasion of mercury takes place. One sees that the angle A of the steep slope roughly amounts to 90 % (which also points to good permeability)

The combination of all the capillary pressures in the tubes results in "the capillary pressure curve". Our example is a "regular" curve because the radii of the tubes increase regularly from left to right. One can also see that at height x only the smallest capillary is filled (e.g. "saturated") with water while all the other capillaries carry oil. Thus the capillary pressure curve of a reservoir rock also indicates schematically how much of the pore space is filled with oil and how much with water (see right side of Fig. 44).

Naturally the arangement of pore radii in a reservoir is not as regular as in illustration 44. More often one would find small pore

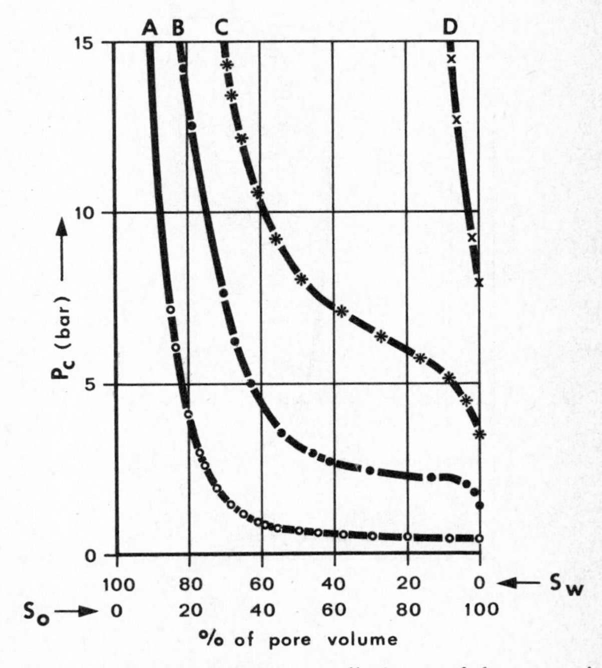

Fig. 46 P_c-curves give a schematic overall picture of the proportions of water and oil in the pore space. At P_c = 10 bar the sandstone C has a water saturation S_w = 60 % and an oil saturation S_o = 40 %. The rectangle as a whole represents the oil + water content of the reservoir. The P_c-curve divides this into water (left side of the curve) and oil (right side of the curve). Sandstone A for instance contains little water but much oil, while sandstone D would contain practically only water with few oil traces. P_c-curves not only give an indication of the saturation but also of the rock's permeability (see Fig. 49)

radii lying next to large ones and vice versa. However, if one were to arrange this randomly distributed tube system according to its pore radii, as the case in illustration 44 shows (and as is automatically the case with the Hg capillary pressure measurement which first fills the smaller then the larger pore spaces with mercury), the outcome would exhibit a very impressive, although schematic picture of the proportions of water and oil in the reservoir pore spaces (or alternatively water and gas, or gas oil and water).

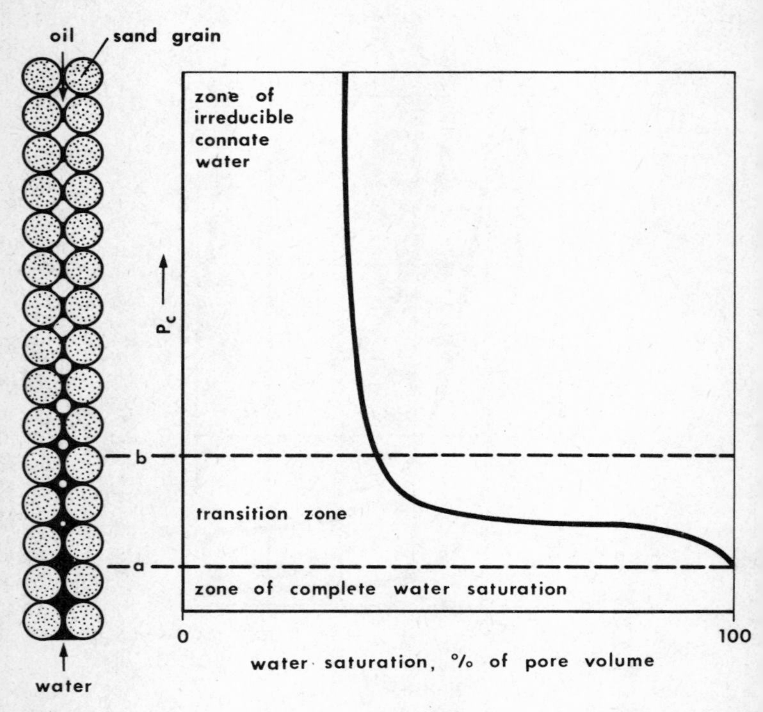

Fig. 47 Each oil reservoir contains, right up to its highest structural level, water that has risen through capillary action. This we call connate water or irreducible water (S_{cw} or S_{wi}). It is found in the tiny spaces at the contact between the sand grains (pendular saturation). Under the connate water zone a transition zone is to be found. In this zone the water covers practically the whole surface area of the sandstone (funicular saturation) and the oil is present only as drops in this water (insular saturation). Beneath this lies the zone of complete water saturation in which the pore spaces are fully water-filled

From the capillary pressure measurement, we therefore see that the old picture of gas, oil and water, separating in the reservoir according to their specific weights along horizontal planes of separation, is no longer valid. This also means that there exists no "edge water line" in its old sense.
In illustration 44 it is clearly visible that, in our container, the zone of 100 % water saturation (that is what in practice is called the "edge water line") lies above the "free water table". The 100 % pore space water saturation zone means the oil-water contact in the pores with the largest radius. How high this zone of 100 % water saturation lies over the free water table depends upon the displacement pressure P_{cd}. In well-sorted permeable pay rocks it lies only slightly higher, but in poorly-sorted reservoir rocks with low permeability, it lies much higher than the free water table. That means that the "oil-water contact or gas-water contact" (that is the level below which all pores are completely filled with water) only runs horizontally when the pay rock is extremely homogeneous and uniformly permeable (see Fig. 47).

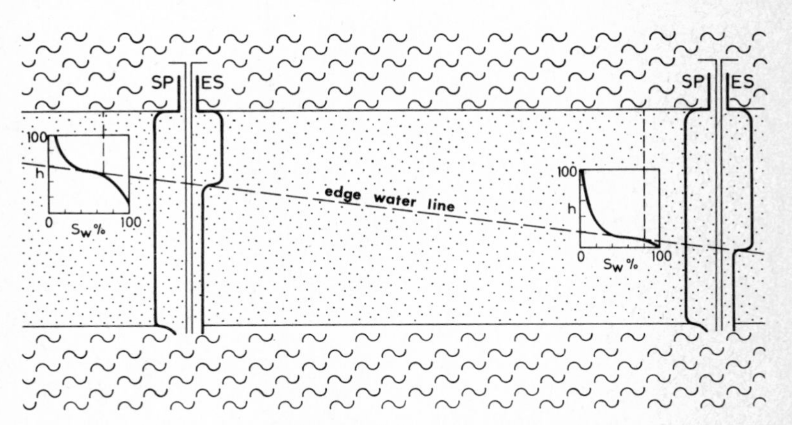

Fig. 48 In inhomogeneous (texturally irregular) pay rocks, the oil water contact lies higher in the fine-pored than in the coarse-pored sections (after *Knutson*[14])

2.3.1.4 *More on P_c-Curves*

Permeability:
From the shape of the P_c-curve certain empirical conclusions can be drawn with regard to permeability. Fig. 45 shows and describes an "ideal" P_c-curve whilst in Fig. 46 P_c-curves from four sands are

PLATEAU AND STEEP SLOPE

well marked		not well marked
(regular pore configuration) (= many equal pores)		(irregular pore configuration) (= mixture of larger and smaller pores)
PLATEAU		
low	high	
(with low P_c)	(with higher P_c)	
(larger pores)	(smaller pores)	
good permeability	average permeability	low permeability

Fig. 49 From the form of the capillary pressure curve, certain empirical conclusions can be drawn with regard to the permeability. "Plateau" and "Steep Slope" are either well-marked (sand 6 of Fig. 50) or not well-marked (sand 1 of Fig. 50). In the latter case, permeability is low. If there is a pronounced low plateau, we may expect a high permeability, while for a high plateau (sand 4 of Fig. 50), permeability will be average (Personal information by Prof. Thomeer)

presented of which sand D, showing a high displacement pressure and no plateau, has the poorest permeability. P_c-curves of seven reservoir rocks of German well shown in Fig. 50, demonstrate the relationship between P_c-curves and permeability.

Pore Radius:
When we consider the pores as capillary cavities and when we take into account that the capillary rise of a wetting fluid increases with decreasing pore radius, we can get an idea of the "pore radius" (or its equivalent) by the following formula

$$r = \frac{2\,\sigma \cos \Theta}{P_c}$$

The fact that the surface tension σ and the angle of contact Θ are well-known physical values, makes it possible to figure out what the pore radius r is if the capillary pressure P_c is known. Figures 51 and 31 show examples.

2.3.1.5 *Determination of S_w, S_o and S_g from P_c-curves*

If one has a capillary pressure curve from the mercury injection method, it is possible to determine the water, oil or gas saturation of a reservoir rock at a given depth. This is shown on a gas containing pay rock (Fig. 52). At first the mercury/air system must be transformed into the water/gas system, and this is done with the help of the physical transformation factor which amounts to 5.1

(see Appendix F). With the help of the specific weight of water and gas one can now determine the height of the capillary water rise. From these one can see, for example, that a capillary pressure P_c of 50 bar in a Hg/air/rock system means the same as 3.65 bar in a water/gas/rock system. This capillary pressure means a capillary rise of 37.5 m of water over the free water table. For such a capillary pressure at this structural depth of the reservoir there exists a saturation of about 35 % water and 65 % gas.

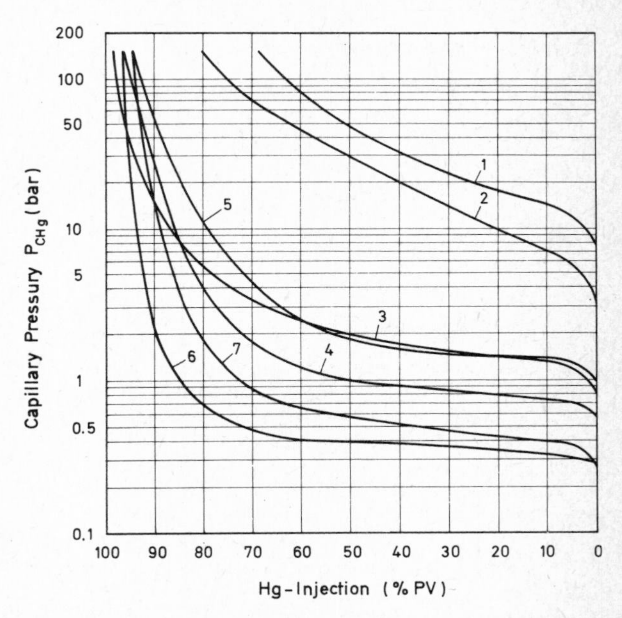

Fig. 50 Capillary Pressure Curves of pay rocks from NW German Wells. (After *Gaida* et al.[10])

Nr.	Formation	Porosity %	Permeability md
1	Upper Carboniferous	8.4	0.14
2	Lower Permian	20.4	1.5
3	Rhetian	11.9	41
4	Lower Liassic	23.9	547
5	Lower Dogger	20.3	101
6	Lower Dogger	22.7	3,550
7	Valanginian	18.3	1.210

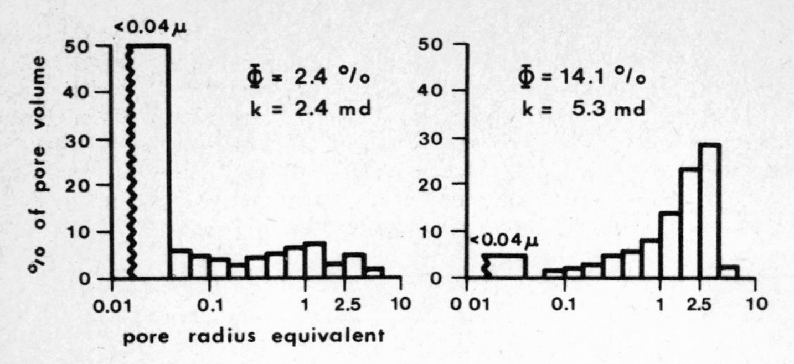

Fig. 51 This illustration shows on the left side an example from the shaly sandstone of the Dogger (a poor payrock with a porosity of 24 % and a permeability of only 2.4 md). One sees that most of the pores have a radius of less than 0.04. (This sandstone is very suitable for a frac-treatment). The right graph shows the Zechstein's Main-Dolomite, a better payrock, with a porosity of 14 % and a permeability of 5.3 md. The dolomite shows a good frequency distribution of the pore radius' equivalent with a maximum of 2.5. (After *Rieckmann*[18])

2.3.1.6 Oil Migration and Capillary Pressure

Capillary pressures certainly play an important role in oil migration, and in nature sudden changes from small to large pore sizes (and vice versa) are more frequent than gradual changes. So, for example, in the case of the change from a shale to a sandstone (the shale being fine-pored, r_t is smaller and P_c larger than in the sand). The difference between $P_{c\ shale}$ and $P_{c\ sand}$ is undoubtedly a motive force.

When dealing with the changeover from sand to shale, the effect is just the opposite; the flow resistance is termed the "Jamin-effect". This flow resistance is commonly so great that the shale overlying the sandstone is designated as a "cap rock" for the reservoir (see Fig. 54).

sediment	r	water/oil P_c	water/gas[6] P_c
shale	10^{-5} cm	5 bar	14 bar
sand	10^{-3}	0,05	0,14

r $= 1/4$ grain radius (by experience)
$\Theta = 0°$
$\sigma_{wo} = 25$ dyn/cm
$\sigma_{wg} = 70$ dyn/cm

Theoretically it may be calculated when the cap rock is sealing. This is the case when the displacement pressure of the shaly cap rock, P_{dc} is greater than P_c, the capillary pressure of the oil reservoir at the structurally highest point. If P_{dc} is smaller than P_c, then the oil migrates further into the overlying layers. The same principle also applies in the case of oil migration into faulted blocks.

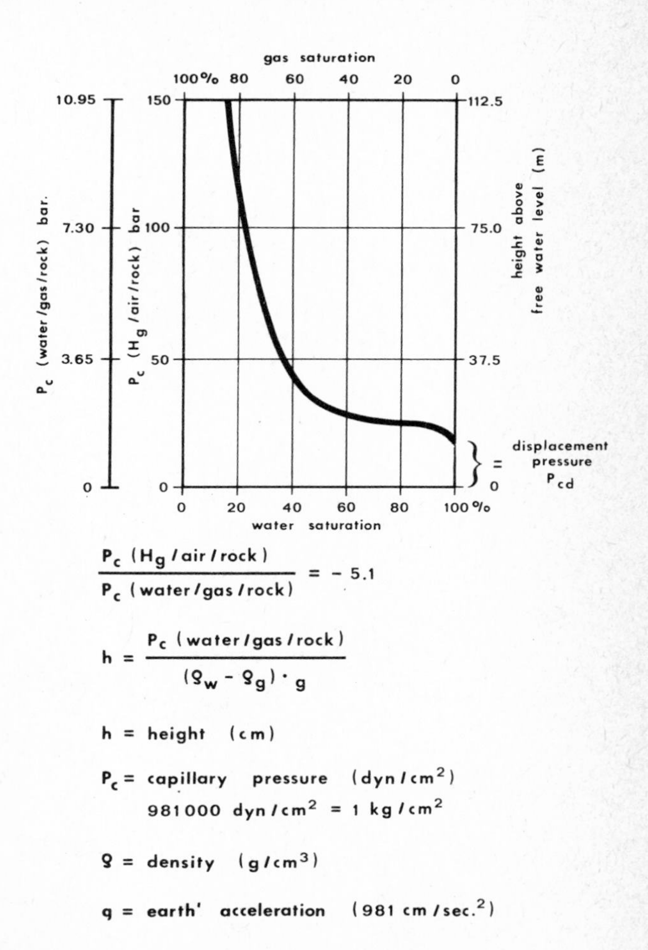

Fig. 52 Transformation of a Hg/air/rock system into a water/gas/rock system and of the P_c capillary pressure into height above free water level

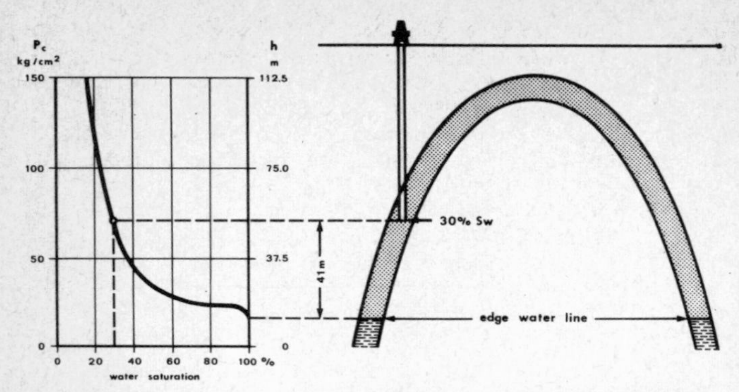

Fig. 53 If, for example, a production test is carried out on the flank of an anticline which yields a 30 % water saturation, one can establish the probable level of the edge water line with the aid of the capillary pressure curve

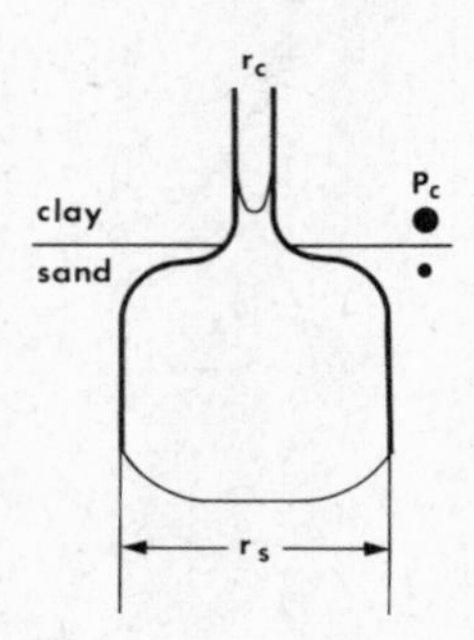

Fig. 54 A sudden change from sand (coarse-pored, small P_c, large pore radius r_s) to clay (fine-pored, high P_c, small pore radius r_c) may explain the sealing function of a shaly "cover" of an oil reservoir

2.3.2 Oil and Natural Gas Under Reservoir Conditions (PVT Relationships)

We are used to seeing oil as it is collected in tanks on the surface. However, oil and natural gas are found under raised temperature and pressure conditions in the pores of reservoir rocks. During the lifetime of a field and in the course of production this reservoir pressure is reduced; with the rising of oil and gas to the earth's surface, the reservoir temperature is also changed.
Crude oil is drawn from the bottom of the drill hole with special sampling instruments. The Pressure-Volume relationships are ana-

lysed in a high pressure autoclave in which the reservoir temperature is maintained. Before coming to the results of the autoclave analysis, it is well to remember the fact that gases are dissolved in liquids (see Fig. 55).

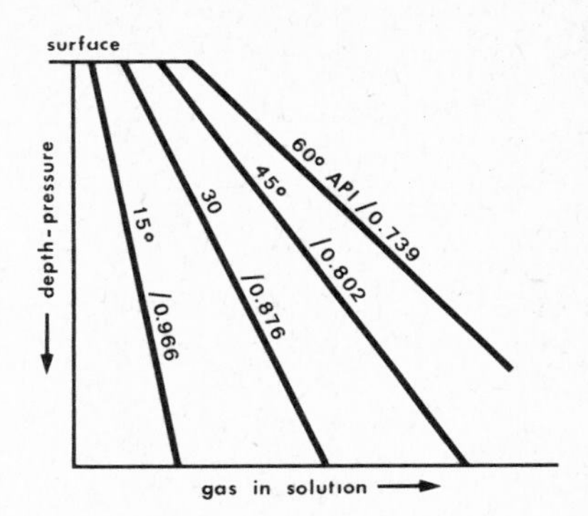

Fig. 55 The solubility of gases in liquids is dependent upon a) the nature of the gas, b) the nature of the fluid, c) the temperature, and d) the pressure

To stick with an oilfield example: the higher the pressure is, the more gas is dissolved in the oil; the lighter the crude oil, the more gas is dissolved (after *Levorsen*[15]). At a certain pressure however, (and a specific temperature) no more gas may be dissolved. This pressure is known as the "saturation pressure".

2.3.2.1 Pressure-Volume Relationships of Crude Oil

A crude oil sample, taken from the bottom of the drill hole, is placed in the autoclave; then the autoclave is put in a water bath whose temperature is the same as that of the actual reservoir. With the help of a piston and with mercury as the liquid one can read off the changing relationship between pressure and volume.

Next, one observes that with a reduction of pressure the volume increases (and vice-versa). This pressure-volume relationship is linear.

At a certain pressure one sees for the first time, through the observation glass, a bubble of free gas. We call this the "bubble-point", and the corresponding pressure "the bubble-point pressure". (This is the same as the "saturation pressure" when dealing with a given quantity of dissolved gas).

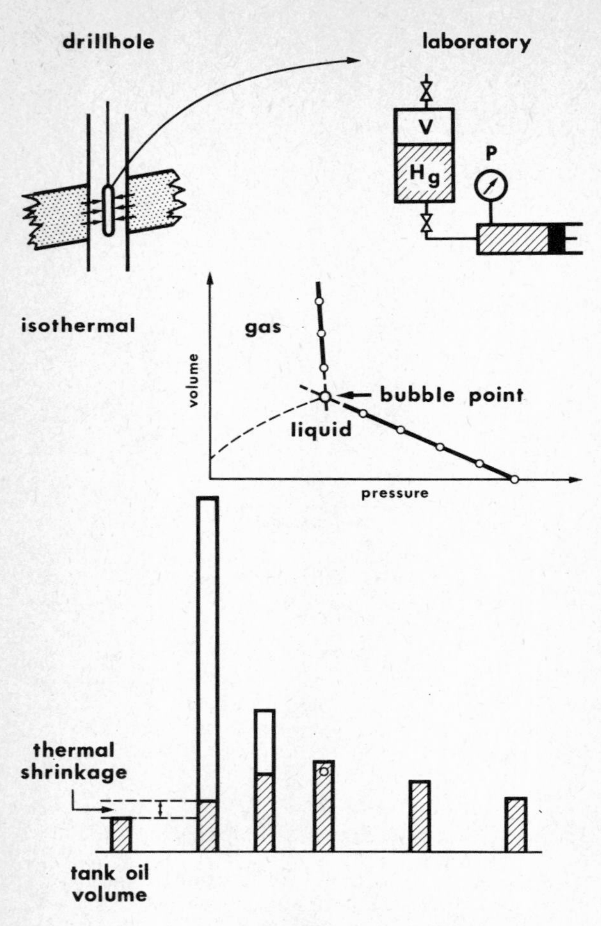

Fig. 56 Schematic presentation of the PVT-investigations and of the PVT-relationships of crude oil

If one lowers the pressure still further, one finds a mixture of free (released) gas and oil in the autoclave. With every reduction of pressure more gas goes out of solution, and the volume of free gas increases sharply in relation to the remaining oil. The volume of fluid decreases further; the oil shrinks, so to speak. (This volume decrease is no longer linear however.) By lowering the pressure to 0 bar the temperature of the reservoir-temperature may be lowered

to the surface temperature. This temperature reduction also produces a further shrinking of the oil, called a "thermal shrinkage". What is left over is the degassed "stock tank oil".
As to definitions, which are especially needed for the calculations, the following is to be added (see Fig. 57):

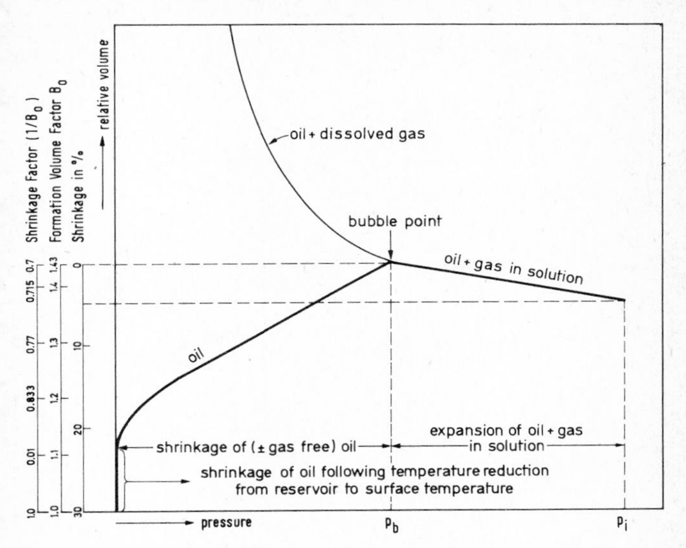

Fig. 57 Terms and letter symbols for PVT-relations of crude oil

a) One can express the "shrinkage" as a percentage. As the largest volume of liquid is reached at the bubble-point, one, as a rule, sets the shrinkage at that point equal to 0. One then sees that the liquid phase at surface conditions has shrunk some 30 % in relation to the bubble-point volume.
b) Since our calculations are always more meaningful when derived from tank oil (the volume being measured at the surface), the so called "Formation Volume Factor B_0" at tank oil conditions is indicated as 1. One can thus read the relative volume of the sample under all kinds of temperature and pressure conditions, (including reservoir conditions), and use it for one's calculations.
c) Shrinkage factor
The "Shrinkage factor" is the reciprocal value of the Formation Volume Factor, i.e. $\frac{1}{B_0}$: Here too the tank oil volume is indicated as 1.

$$Bo = \frac{\text{reservoir oil}}{\text{tank oil}} = > 1$$

$$\frac{1}{B_0}\text{(formerly called C)} = < 1$$

c_0 = compressibility of oil (for example 10×10^{-5}/bar)

2.3.2.2 Pressure-Temperature Relationships (Phase Relations)

The relationship between pressure and temperature gives us information about the aggregate condition of the pore content, and an indication as to whether this pore content is liquid or gaseous. The simplest relationships exist in a single substance such as Ethane. An oil reservoir, however, is not composed of a single substance, but rather consists of a mixture of numerous hydrocarbons. From the "dry gas reservoirs" (with a high methane content) through rich or "wet gas reservoirs", to condensate fields and then to oil fields with high gas-oil ratio and, finally, to undersaturated oil

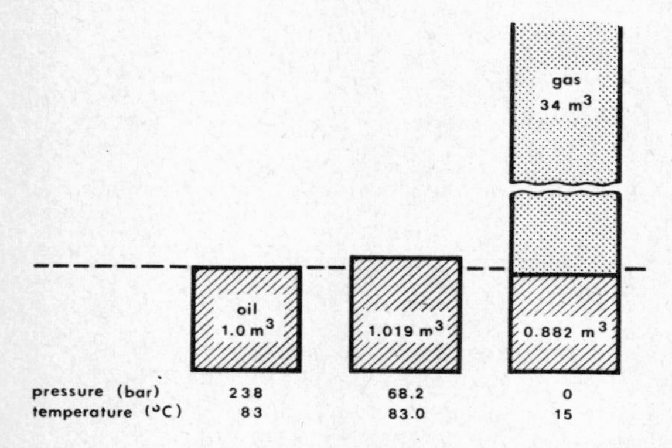

Fig. 58 The crude of the oil field Suderbruch (Germany) is illustrated as a practical example. The original reservoir pressure amounted to 238 bar, and the reservoir temperature was 83 °C. With a reduction of the reservoir pressure, but at the same temperature, the liquid phase expands, and this continues up to 1,019 m³, at a pressure of 68.2 bar. This pressure of 68.2 bar is the "bubble-point pressure" of Suderbruch crude oil. As the pressure sinks further, continually more gas is released; the liquid phase shrinks steadily. The temperature is also reduced to the surface temperature, and at a pressure of 0 bar, a cubic meter of reservoir liquid has shrunk to 0.882 m³ and 34 m³ of free gas has been released. This relationship of 1 : 34 is called the "original gas-oil ratio" with regard to the crude oil

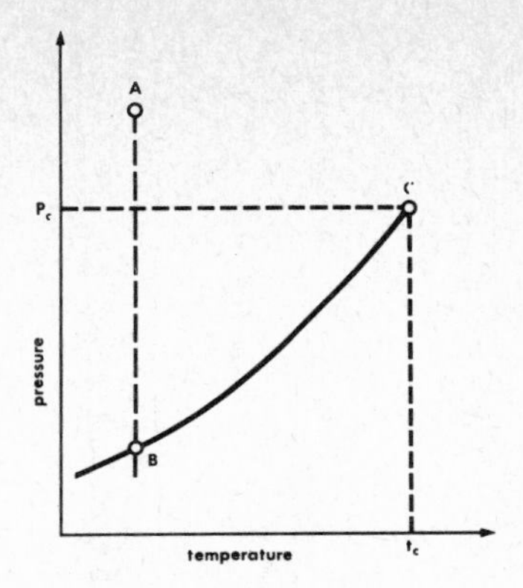

Fig. 59 The simplest relationships exist in a single substance such as Ethane. If reservoir pressure and temperature correspond to point A, then the Ethane would be found, in the pores of the payrock, as a liquid. Should the reservoir pressure descend along the dotted line (isothermal pressure descent) then we would first detect gas at point B. With continued expansion more gas would develop at constant pressure. Only after no more liquid were present would the pressure diminish further. The connection of all B-points results, as we know, in the vapour-pressure-curve of Ethane. Below the vapour-pressure-curve the Ethane is gaseous, above the curve it is liquid. Above point C, the critical point, two phases can co-exist

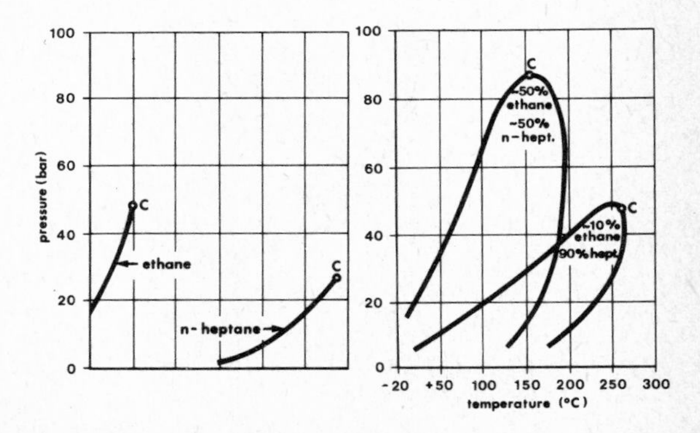

Fig. 60 PT-Diagram for the ethane-n heptane system. The connection of all C-points is called the "critical envelope curve" of the system. (After *Pirson*[17])

reservoirs (with low gas-oil ratio), all transitions are possible and known.
Instead of the simple vapour pressure curve of Ethane, we now see a phase diagram whose form is totally dependent upon the composition of the mixture of hydrocarbons. For example, the more plentiful the hydrocarbons, the more bulbous the diagram becomes, (and the higher the critical point C advances). Within the dark bulbous outline, gas and liquid can coexist, (within the percentage outlined in Fig. 62); outside the bulging line only one phase exists: above the bubble-point line there is only liquid, and below the dew-point line only gas.
During production, as a rule, only the reservoir pressure decreases, while the temperature stays more or less the same. Thus the phase relations of a crude oil may be read from the vertical line indicated. Only at the surface is there a fall in temperature to the surface temperature as indicated by an X in Fig. 62 and 63.

2.3.2.3 *Viscosity of Crude Oil*

According to Dacry's lawe, (page 19), viscosity is the physical property of an oil which has the greatest influence on its flow within the rock reservoir. Viscosity of crude oil is known to be lowered by temperature. At reservoir temperatures and mid – depth conditions the viscosity of crude oil is often only half of what it

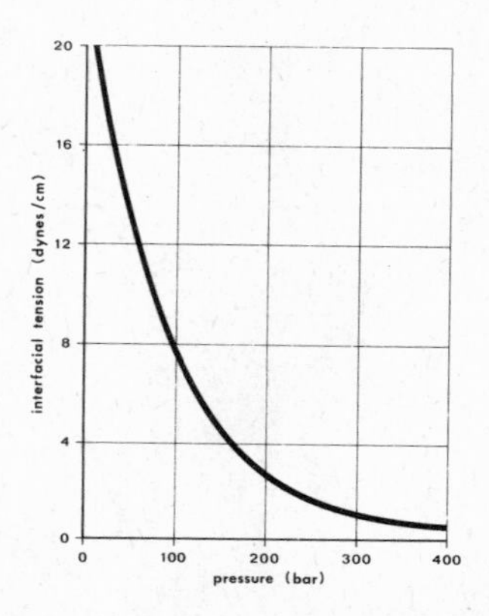

Fig. 61 In the phase behaviour the "Critical Point" C represents the point at which the liquid and vapour have identical qualities: all physical properties of the phases merge. This is impressively shown by the manner, in which the interfacial tension between oil and gas decreases at high pressures. The above example is taken from samples of Iranian fields. (After *Saidi*[21])

is at the surface. Gas released from an oil reduces its viscosity; rising reservoir pressure raises it.

2.3.2.4 Super Compressibility of Natural Gas

In an ideal gas the product of the pressure and volume is constant ($100\ m^3$ of an ideal gas under atmospheric pressure, at a pressure of 100 bar absorbs $1\ m^3$ of volume). Expressed mathematically

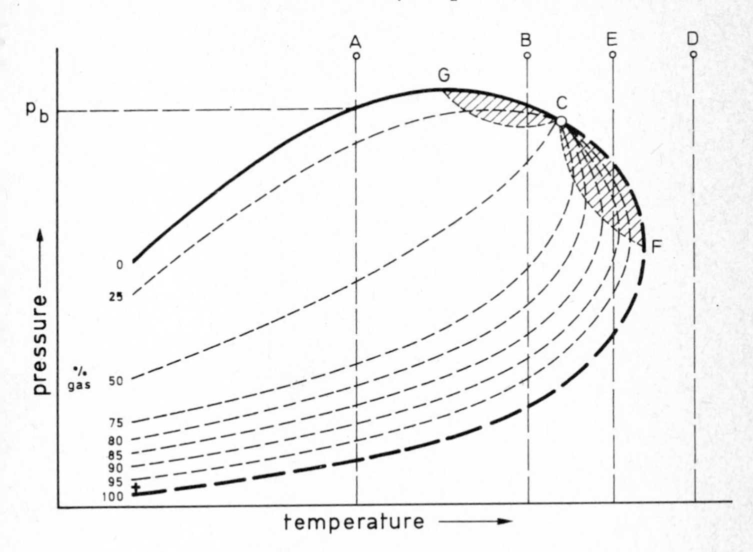

Fig. 62 Let us assume that we take an oil sample whose pressure and temperature correspond to point A. If, in an autoclave, pressure is reduced along the dotted line while temperature remains constant, the first gas bubble will appear when reaching the bubble point pressure (p_b). When pressure is further reduced, the proportion of gas will increase more and more until, after reaching the "dew point line", the total reservoir content has turned to gas. In principle, the same is true for higher reservoir temperatures, as at point B. In the cross-hatched areas "retrograde condensation" will occur (revaporization of the liquid phase and recondensation of the gaseous phase), which is an important phenomenon in the production from condensate fields. Point F indicates the maximum temperature at which a liquid may be condensed from a vapour (= cricodentherm; accordingly point G is called crivapourbar). C represents the point at which the liquid and vapour have identical qualities, e.g. all physical properties of the phase merge. For a sample at point D, a change of the gaseous phase would only occur after reducing the temperature to surface temperature (point X). (After *Schmid*[22])

$$PV = nRT$$

where

P = Pressure
V = Volume
n = number of moles
R = gas constant (proportionality constant)
T = temperature

Natural gas, as it occurs in nature, is not an "ideal gas" but is composed of a mixture of various hydrocarbons plus "invaders" of N, CO_2 etc. Therefore the designated product is no longer consistent as far as pressure and volume are concerned. Rather it varies according to temperature, usually by a factor known as the Z factor.

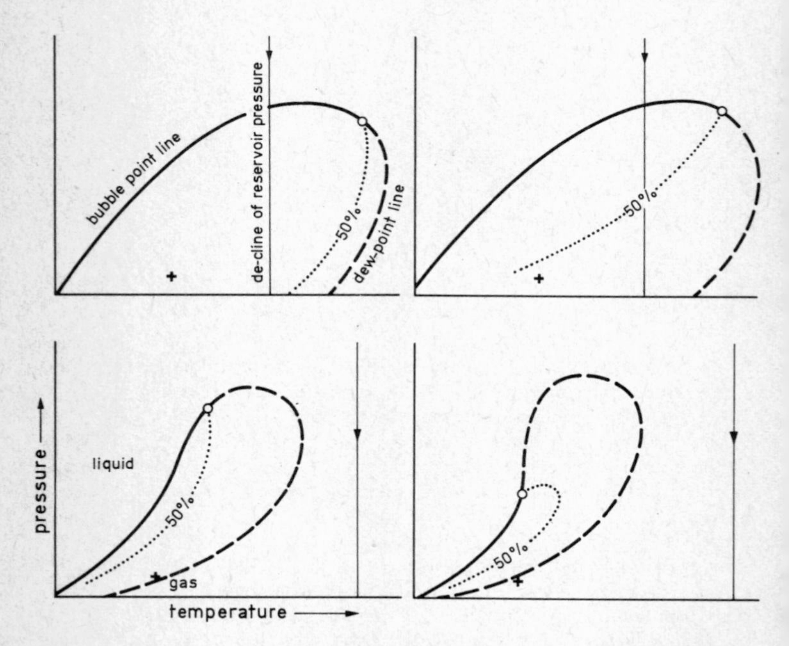

Fig. 63 Phase diagram of crude oil with low (upper left) and high (upper right) gas content. A "wet gas reservoir" is characterized by the phase diagram at lower left, a "dry gas reservoir" by that at lower right. The arrow indicates the fall in reservoir pressure, the cross marks the temperature at the surface. (After *Clark*[4])

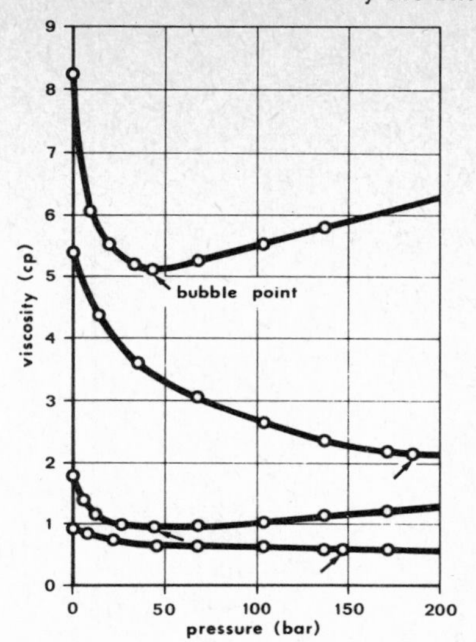

Fig. 64 Shows the viscosity for various crude oils under reservoir conditions, as reservoir pressure diminishes and gas begins to be released. We see, at first, that as the reservoir pressure is reduced, the viscosity diminishes; however beneath the saturation pressure it increases due to the release of gas. (After *Pirson*[17])

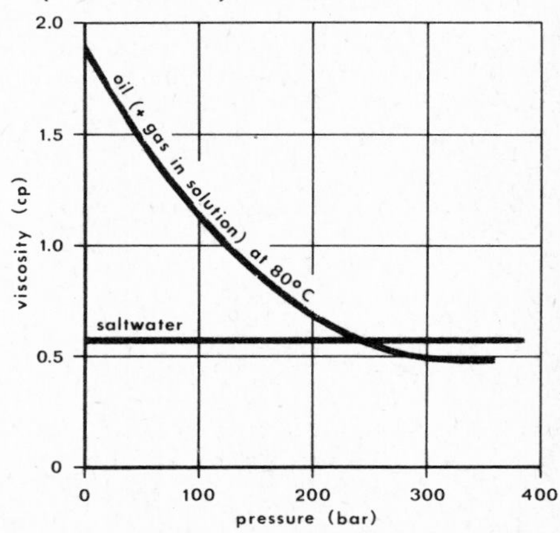

Fig. 65 The relationship between the viscosities of crude oil and edge water is of special importance because of the formation of water tongues and water cones; also in secondary projects, i.e. flooding. The viscosity of crude oil can diminish to as low a viscosity as that of salt water and crude oil and can achieve the same mobility as edge water through this process

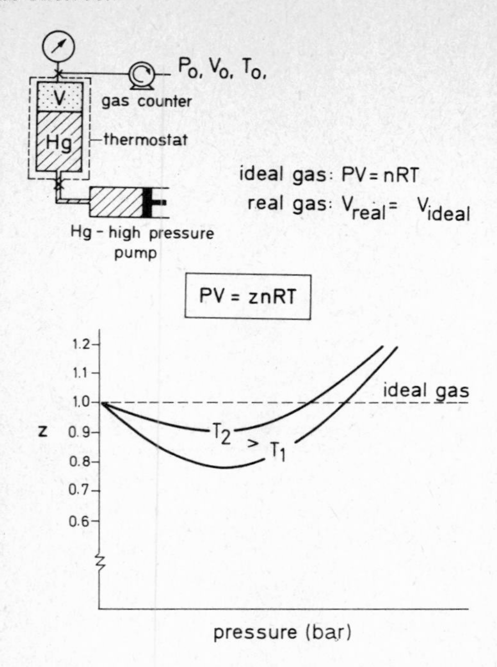

Fig. 66 Shows that at a temperature of T_1 the compression of a natural gas is less than that of an ideal gas; Z declines to about 0.8. Only after a certain pressure is reached does the compression of natural gas become greater than that of ideal gas. Z rises quickly above 1.0. Since, as stated, this Z factor is dependent upon temperature, it has to be determined anew, in the autoclave for each and every natural gas. Only with the help of this Z factor can one work out the supply and material balance determinations. (For the formation-volume factor B_g, see Appendix G

2.3.2.5 Sampling

Special devices are used for sampling. These instruments contain a chamber (cubicle) for the intake of oil samples and are lowered into the borehole. Two main systems may be outlined:

a) The Flush System (Fig. 67)
b) The so-called In-flow System (Fig. 68)

Naturally the temperature at the sampling depth is also taken. This is the reservoir temperature, since the sample is normally taken at the oilsand level. The autoclave research is then carried out at this temperature.

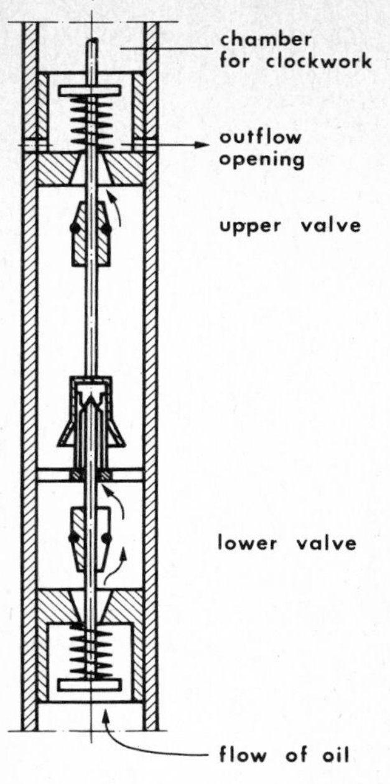

Fig. 67 Sampling device – Flush System. The oil coming out of the perforations of a producing well flows through the aperture, while an upper and a lower valve are open. A built-in timing mechanism closes after a given time both valves (when it has been ascertained that the chamber has been completely flushed)

Fig. 68 Sampling device – Inflow System. The oil flow enters the chamber at a point, where through an ingenious construction, (namely a valve pipe which is also moved by a timing device), the chamber is filled with oil at reservoir conditions and then shut

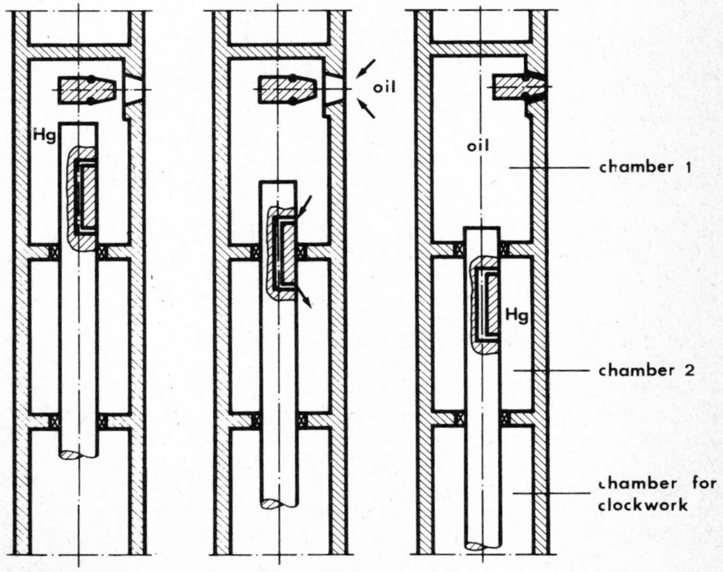

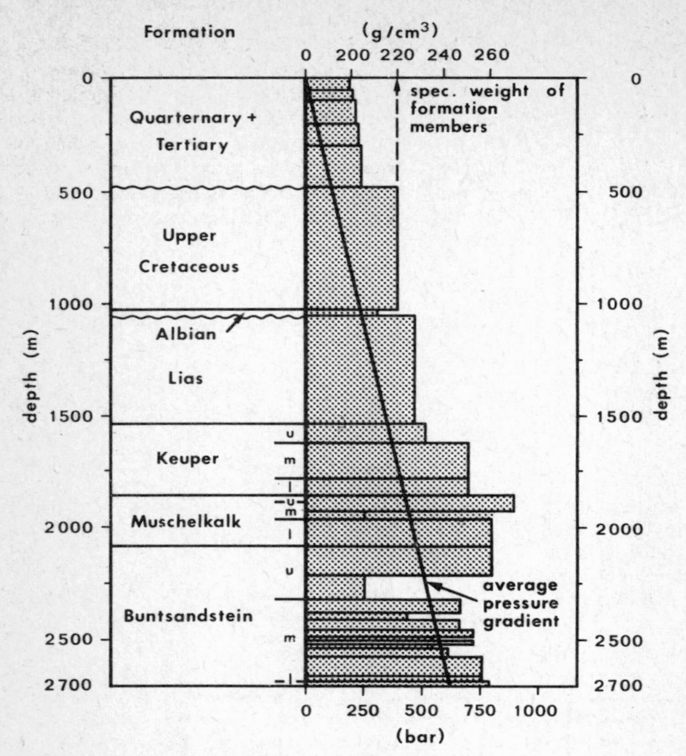

Fig. 69 Usually a well drills through formations with low (e.g. salt) and high (e.g. limestone) specific weight. The petrostatic pressure is the product of the specific weight of the formation members and their thickness. At the total depth of 2700 m of this German well, the petrostatic pressure is ~ 650 bar. The average pressure gradient is ~ 2.4 bar/10 m (and therefore corresponds to an average specific weight of the rocks of 2.4 g/cm³)

2.4 Reservoir Pressure and Reservoir Temperature

The pore content of the payrock is always under a certain pressure; we call this the "reservoir (fluid) pressure". The payrock as such is usually under much higher pressure. We call this the "petrostatic pressure". It is commonly brought about by the pressure of the overlying rock layers (overburden pressure), but at times also through tectonic pressures. Inside the payrock, the relationship between petrostatic and reservoir pressure is called "grain pressure".

a) Petrostatic Pressure
The petrostatic pressure is normally caused by the pressure of overlying rock layers. As a rule, it depends on the depth and specific weight of the overlying rocks.

$$p_{petrost.} = \frac{\varrho_r \cdot D}{10} \text{ (bar), where } \varrho_r = \text{specific weight of overlying rocks}$$
$$D = \text{depth (m)}$$

The "pressure gradient" (rise in pressure per meter of depth) therefore depends completely upon the specific weight of the rocks and averages usually 0.23–0.24 bar/m (see Fig. 69).

b) Reservoir Pressure
The pressure, to which the pore content of the pay rocks is subjected, is not dependent upon petrostatic pressure. Usually it is the same as the hydrostatic pressure of an equivalent column of salt water

$$p_{hydr.} = \frac{\varrho_w \cdot h}{10} \text{ bar where } \varrho_w = \text{specific weight of water}$$
$$h = \text{height of water column (m)}$$

The pressure gradient, therefore, normally is 0.10–0.13 bar/m. This is to be attributed to the fact that the payrock normally crops out somewhere at the earth's surface, and thus the water-filled pores act as a hydrostatic column. We shall see below that there are many exceptions to this norm.
Since in an oil or gas reservoir the pressure depends upon the specific weight of the reservoir contents, the reservoir pressure, especially in natural gas-fields with a high gas column, differs greatly from the hydrostatic pressure (Fig. 70). It must be carefully

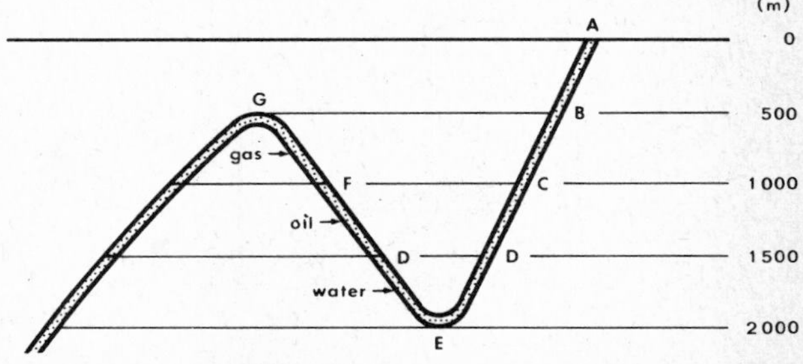

Fig. 70 If water has a specific weight of 1.1, oil of 0.8 and gas of 0.2 g/cm³, then the pressure at D (oil/water contact) is 165 bar, the pressure at F (gas/oil contact) 125 bar. At the same depth, but in the water-filled part at C, it is only 110 bar. At G, the top of the structure, the pressure is relatively high: 115 bar compared to 55 bar at B

estimated before the drilling procedure and the casing design are determined. For this reason, especially in steeply dipping gas-filled reservoirs where extremely high pressures may occur, the drilling staff must be warned in good time (see Fig. 71).
It may also be explained from these simple physical laws that, besides higher hydrostatic pressures, lowered pressures may also occur, depending on the piezometric level (see Fig. 72).
From this rule, namely the dependence of the reservoir pressure upon the depth and specific weight of the pore content, large

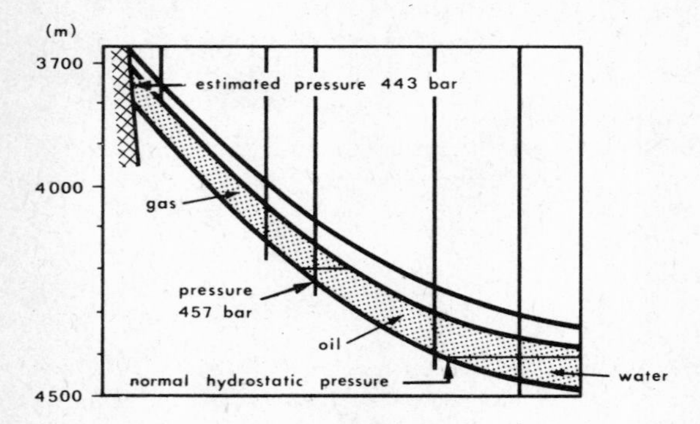

Fig. 71 In steeply dipping payrocks with a thick gas column, the reservoir pressure in the structurally highest part is extremely high. It is the petroleum engineer's special task to warn the drilling people against such areas of extreme pressure when designing the casing scheme

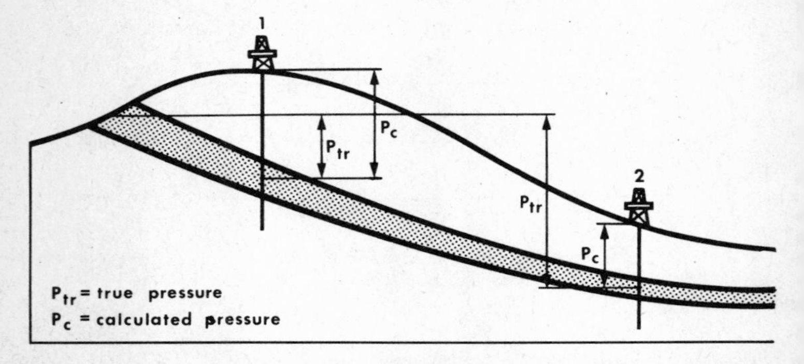

Fig. 72 It may also be explained from simple physical laws that, besides higher hydrostatic pressures, lower pressures may also appear, depending on the piezometric level. The true pressure (P_{tr}) in this Figure differs considerably from the pressure (P_c) calculated from the depth of the well

upward variations are possible. These irregularities were first investigated along the Gulf Coast of the U.S.A., where in thick shale-sand sediments high pressure zones were encountered, which came near the petrostatic pressure. These high pressures occurred from a known depth and were called "geopressures" in order to distinguish them from "hydropressures". In any case these observations may not be generalized. They appear to be restricted to monotonous sand – shale series in large sedimentary basins.
As shale is compacted, the originally great porosity diminishes relatively fast (see page 10). In general, at this point of the sedimentary process, trapped pore-water escapes in proportion to the degree of compaction. If the permeability decreases faster than the porosity, the rock becomes impermeable before all of the sedimentary water can escape. The same effect is produced if sedimentation, and thus overburdening of additional sediments, occurs extremely rapidly. If this happens, the pore contents are placed

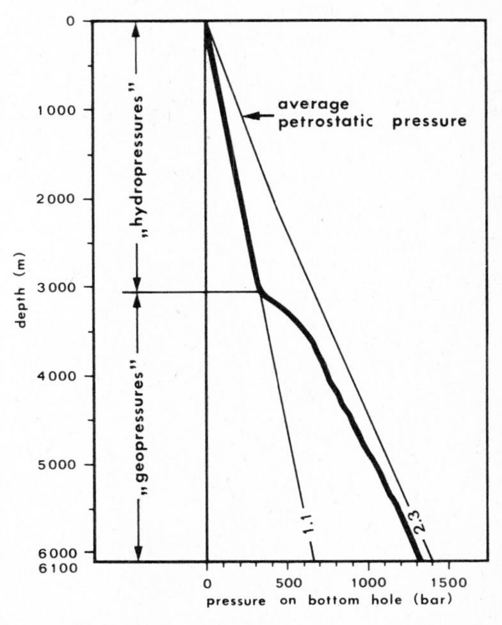

Fig. 73 In large sedimentary basins with thick ± homogeneous sedimentation there exists an overpressure down from a certain depth, which is greater than the hydrostatic (gradient approx. 1.1 bar/10 m) and almost equal to the petrostatic (gradient ~ 2.3 bar/10 m) pressure. These high pressures are called "geopressures", unlike the normal "hydropressures"

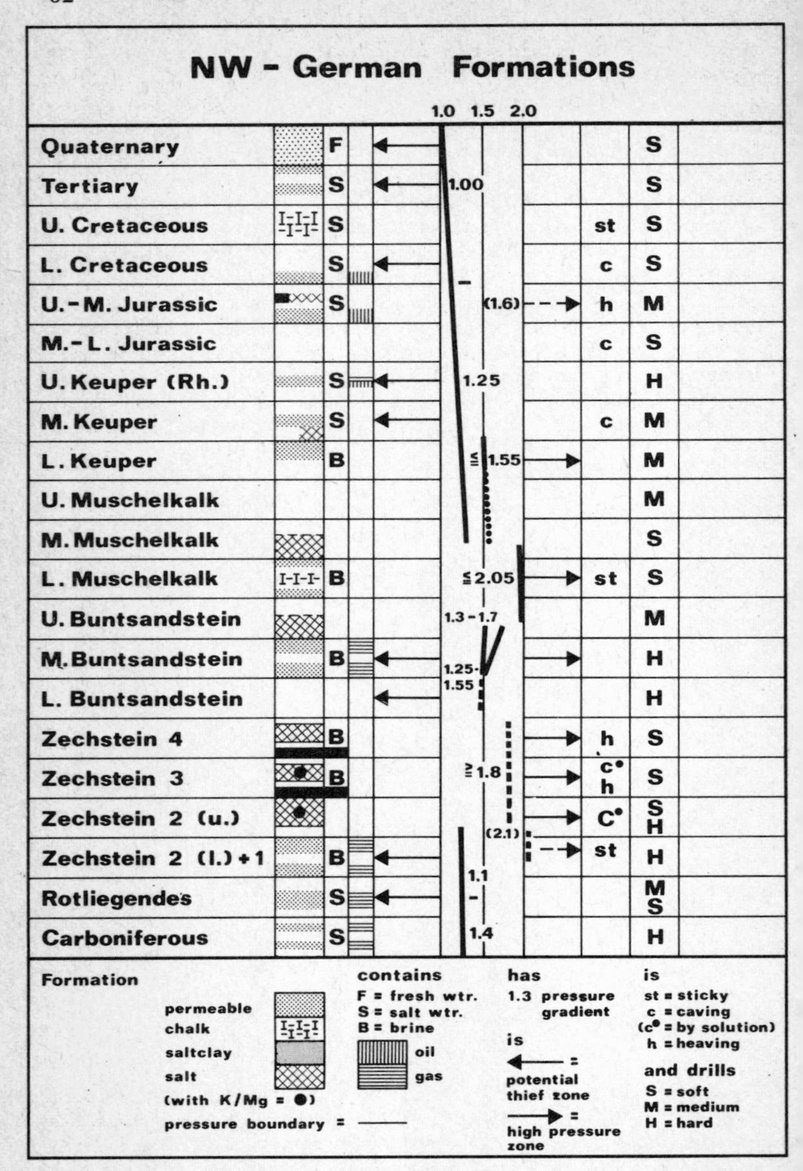

Fig. 74 (By courtesy of Dr. *D. Betz*)

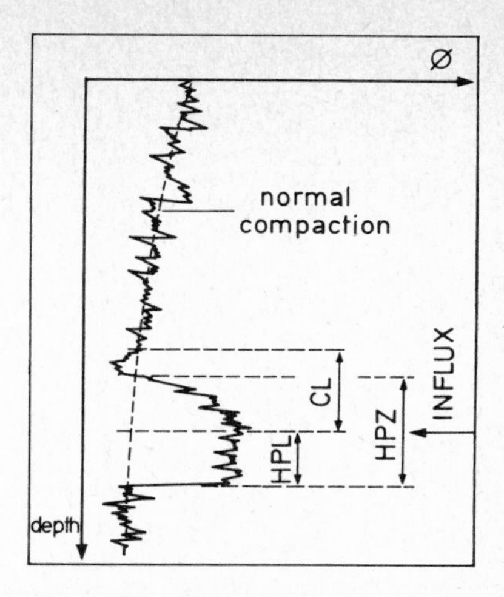

Fig. 75 A High Pressure Zone (HPZ), composed of a "covering layer" (CL) and a "high pressure layer" (HPL) is clearly indicated by an increased porosity. The HPZ is marked by an influx of water. (After *Rizzi*[19])

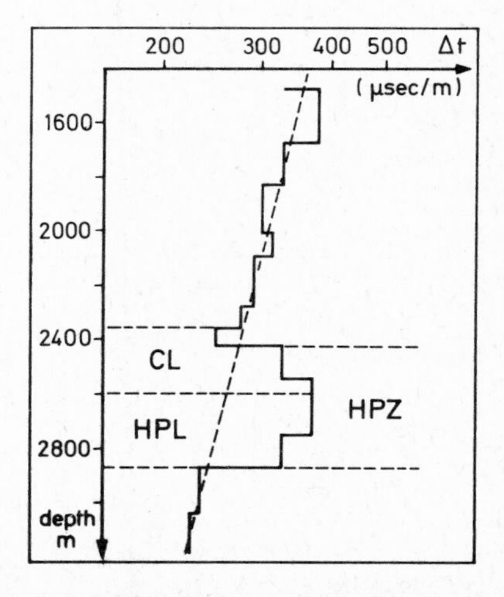

Fig. 76 The Formation Travel Times of seismic surveys clearly indicate High Pressure Zones (HPZ). (After *Rizzi*[19])

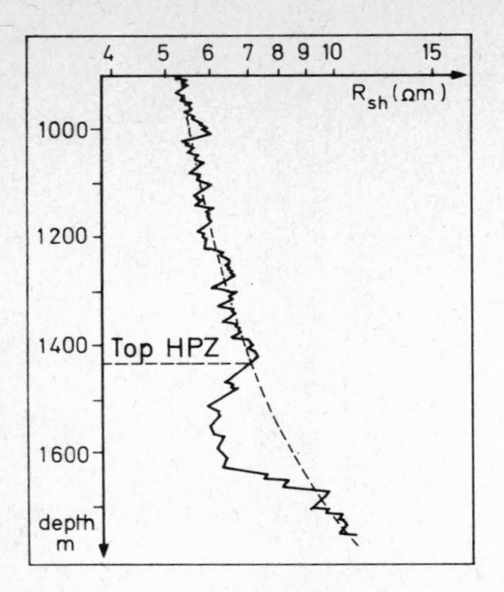

Fig. 77 In High Pressure Zones (HPZ) the electrical resistance of shale shows a well – marked decrease. (After *Rizzi*[19])

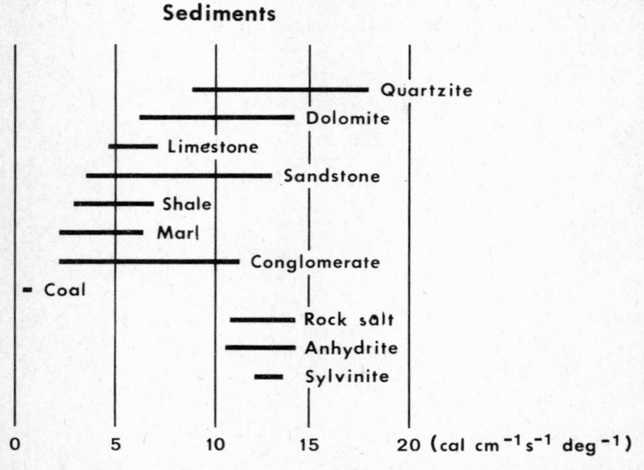

Fig. 78 Thermal conductivity of various sedimentary rocks. (After *Kappelmeyer*[12])

under continually greater petrostatic pressure, which creates a number of drilling difficulties.

If there is, for example, a sand lens in a shale of this type with no surface connection, then the pore content of the sand lens would understandably be under an equally high reservoir pressure. Even "fossil pressures" may occur, for instance in some shallow oil-fields of Northern Germany situated below sealing shales, which were covered by a thick ice layer in the Quaternary Period and which are now unloaded after the retreat of the glaciers.

A variation is the capping of payrocks (those which do not crop out at the surface) by other impermeable layers, especially salt. With this combination it should be expected that the reservoir contents of the payrock will come under increased pressure, and at times petrostatic values may be reached. Outside of large sedi-

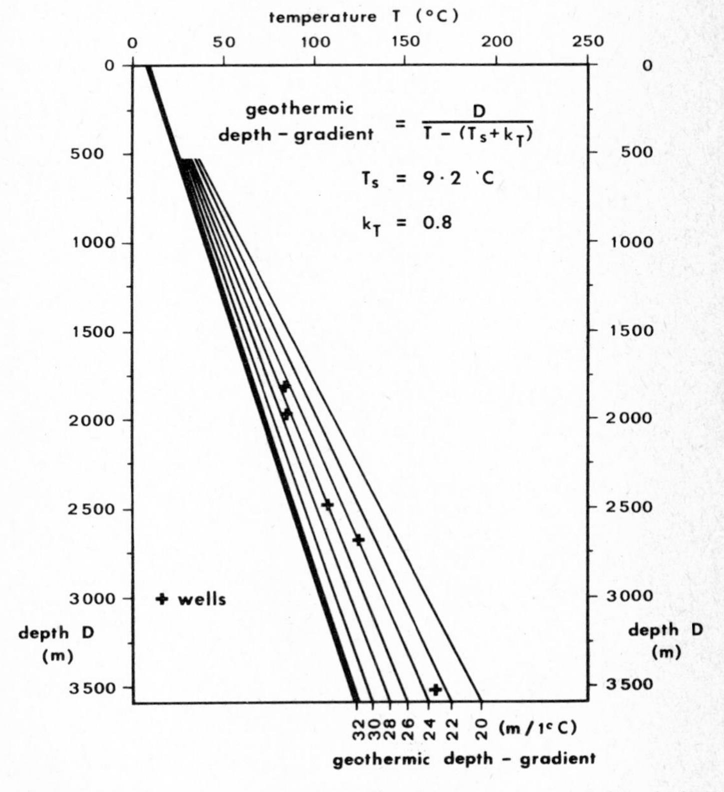

Fig. 79 Temperature in several wells of a German gas-field which is marked by a somewhat reduced geothermic depth gradient. (D = depth in m; T = temperature in °C; T_s = average annual surface temperature in °C; k_T = geographical constant)

mentary basins with monotonous shale-sand series, complicated phenomena should therefore be expected which may not be expressed by a simple definition. Investigations relating to this were made in North German payrocks by Dr. Betz, and it may easily be seen from Fig. 74 that extremely high pressure gradients are always found under impermeable salt layers.

It would certainly be of great importance to both the drilling and the reservoir engineer if such variations from the normal pressures could be predicted in advance (or at least determined during the drilling procedure). In this respect initial successes have been achieved. These result, in the last analysis, in the observation that raised pressures manifest themselves in

raised porosity,
raised travel times in seismic measurements
lowered electrical resistance.

From the above it may be seen that the pressure to which the pore space is subjected, produces (so to speak) a "counter pressure" against the petrostatic pressure. It relieves, as it were, the pressure which the individual sand grains of a sandstone exert upon one another. This pressure is called "grain pressure", and it is to be expected that this grain pressure increases the lower the reservoir pressure is; that means, normally, the more oil or gas is removed

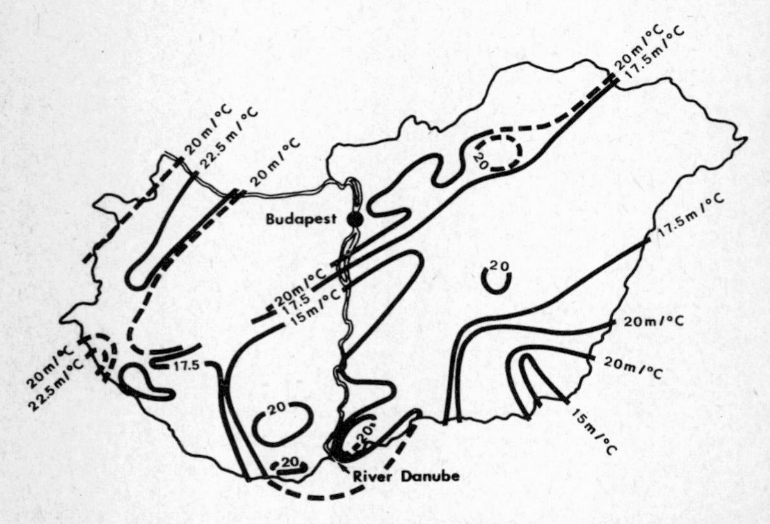

Fig. 80 In Hungary, the "geothermic depth gradient" or "reciprocal temperature gradient" is unusually high. In the F. R. of Germany, the average value is ~ 33 m per 1 °C rise in temperature. (After *Baldizsár*[2])

from the pay rock. This may go so far as to lead to additional pay rock compaction and soil subsidence on the surface (see page 14).

Reservoir Temperature

Apart from higher pressures, oil, gas and water are also subject to higher temperatures in the reservoir than on the surface; these higher temperatures also change their physical properties, compared with their surface state (see page 46).

"Geothermal energy is essentially nuclear energy from the big natural nuclear reactor situated in the crust and mantle of the earth. The nuclear fuels are the K^{40}, Th^{242} and U^{238} atoms dispersed chiefly in acid crustal rocks. Heat flow measurements support the view that the so-called Tertiary or Alpine global orogenic zone is characterized by surface heat flows higher than the average value. On the other hand Archaic shields exhibit low surface heat values. The amount of geothermal heat in the earth is not only immense, but it is well insulated and only a small fraction is conducted from the surface into space. The amount of heat given off by volcanic activity is negligible compared with the amount of conducted heat. It may be that radioactive heat production is even greater than heat loss, and that the earth is actually heating up".[2]

By "geothermic depth level" or "reciprocal temperature gradient" one understands the depth increase per unit rise in temperature. In Germany one uses as an average value 30 to 33 m/1 °C; in the U.S. a somewhat higher value is employed.

The fact that increase in temperature is not only dependent upon heat conductivity of the rocks makes one realize how many variations there are from these normal values: distance from mountain ranges, volcanic masses, gas and water content and other factors cause extraordinary deviations, and therefore there are regional areas with very low geothermal gradients.

In 1920, the first oil-field temperature measurments had already shown that the temperature increase with depth was greater over anticlines than in flat-lying beds or over synclines. In the case of salt domes, the salt – a good heat conductor – originates from great depths where temperatures are high: Higher temperatures are therefore to be expected right above salt domes. This makes it possible to map salt domes from satellites.

The higher temperatures at greater depths may possibly set limits to deep drilling techniques, due to the break-down of high polymers, and the organic compounds in the drilling mud. During the production phase, the reservoir temperature does not change considerably. Naturally, flooding with hot or cold water, injection of steam and all other secondary geothermal measures do have some effect.

2.5 Reservoir Energy (the driving forces)

The flow of oil through the reservoir rock to the well is caused by various forces "stored" in the reservoir. These forces may become active either individually or in combination; frequently they become effective one after another. It is uncommon for an individual source of energy to be active alone, and frequently it is not even possible to define clearly the source of energy which, in a given period, is the main factor active in the reservoir. In spite of this partly interdependent and partly independent action of the various sources of energy, it is possible and advisable to classify the reservoirs according to the type of reservoir energy. The driving forces in question are the following:

1. expanding gas
   a) liberated gas
   b) gas cap gas
2. encroaching edge water (or bottom water)
3. expansion of oil + dissolved gas
4. gravity and others.

As soon as a reservoir is struck and produced by a well disturbing its pressure equilibrium, the above forces begin to drain the oil and gas towards the point of lowest pressure which, in the present case, is the well.

2.5.1 Dissolved gas drive

In the widest sense of the word, a gas drive is any energy mechanism governed by the expansion of free gas. Obviously, this applies in the first place to gas-fields. Among oil-fields it applies to those possessing a gas cap as well as those in which the gas is liberated from the oil by pressure release.

As a rule, gas drive is restricted to "closed" ("volumetric") reservoirs. A "volumetric reservoir" may be defined as an oil and/or gas reservoir the volumetric delimitation (or, in other words, the productive area) of which does not change in the course of production. It can, therefore, be compared to a pressurized, heated tank. In other words, there is no edge water encroaching into the reservoir and replacing the volume withdrawn from it by production. This may have several causes. The reservoir may be surrounded by non-permeable rocks or the edge water zone may be separated from the productive part by a tight fault. Even fields the aquifer of which is insignificant as compared with the petroliferous zone may still be said to form a "volumetric reservoir".

In a dissolved gas drive reservoir the start of production and the pressure release in the vicinity of the well cause a liberation of gas. This will either be the case right from the start (when the oil is saturated with gas) or it will be retarded until bubble point pressure

is reached (when the oil is undersaturated). Initially the gas will form small bubbles in the pores as shown in Fig. 81. With increasing relative permeability (k_{rg}) the liberated gas flows either to the well (increase of the gas/oil ratio) or rises to higher parts of the reservoir structure where it then forms a "secondary gas cap".
In such fields, reservoir pressure declines continuously as a function of production. When plotting reservoir pressure against cumulative production the result is an almost linearly declining curve (Fig. 82). When the field is closed, reservoir pressure will not rise.

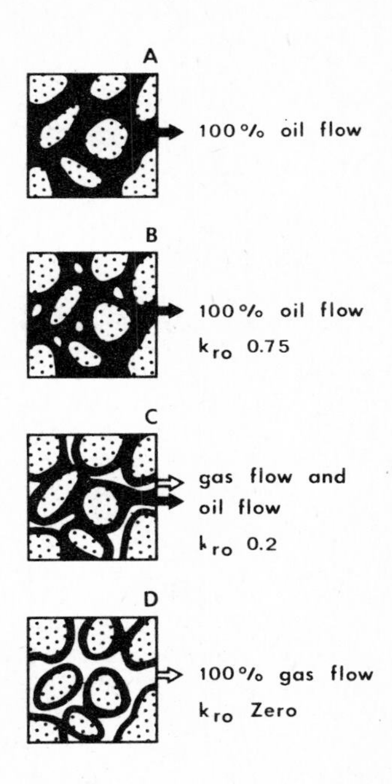

Fig. 81 At high pressures (A) gas is dissolved in the oil (black) occupying the pore space between sand grains (dotted). There would be a 100 % oil flow through the reservoir. The pressure drops by the oil withdrawal. Consequently gas envolves from solution and occurs as small separate bubbles (white). These bubbles reduce the relative permeability to oil k_{ro} to 0.75 (B). Oil flows at reduced rates. With a further pressure drop (C) the size of gas bubbles increases until they join in a continuous thread of gas. Oil and gas flow together. The relative permeability to oil decreases, the gas/oil ratio increases. Finally (D) the relative permeability to oil becomes zero; there is a 100 % gas flow. (After *Clark*[4])

All that may happen is a compensation of pressure between the individual wells. Wells in such fields require early pumping.
In such fields, production, too, shows pronounced characteristics. The first gas liberated will not flow in the beginning so that, initially, the gas/oil ratio may be slightly lower than the solution ratio. When effective permeability sets in and the gas begins to flow the gas/oil ratio rises rapidly up to a certain maximum along with declining pressure and increasing gas liberation. When this maximum has been attained it will drop again until the field is depleted.

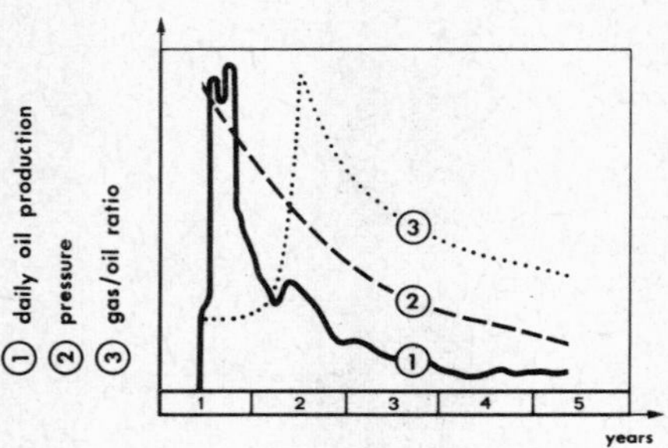

Fig. 82 In a "Dissolved Gas Drive Reservoir" the production rate decreases rapidly after the peak; reservoir pressure declines continuously; GOR rises to a maximum and then drops; no water; pumping at early stage; recovery factor 5–30 % of original oil in place. (After *Clark*[4])

Temporarily closing in wells with high GOR is particularly effective in fields with a steep dip (and, therefore, gravity force).
In dissolved gas drive fields the recovery factor is low. Statistics covering a substantial number of U.S. fields show it to vary between 10 and 60 %. Its average is believed to be 33 % (Fig. 107). According to experience in the Federal Republic of Germany this value is too high for conditions prevailing there, presumably due to the pronounced lack of homogeneity of many reservoir rocks. In the Federal Republic of Germany, recovery in such fields is usually expected to amount to between 5 and 30 % of the oil in place. Due to residual oil saturations of this order, such fields are suited to an application of secondary recovery measures, in particular of water-flooding.

2.5.2 Gas cap drive

In the case of the "liberated gas drive" the expansion of the liberated gas acts, so to say, "from the inside". In the case of the gas cap drive, this energy mechanism acts "from the outside". In frequent cases, though, a "secondary gas cap" will form in the course of production and cause the first type to gradually change over into the second. This will happen in particular

a) when the structural characteristics are pronounced so that there is a steep dip
b) when the reservoir rock is of substantial thickness
c) when production is so slow that the force of gravity can intervene and bring about a separation of oil and gas
d) when the permeability of the rock is high and, finally,
e) when the viscosity of the oil is low.

However, the formation of a "secondary gas cap" is also favourably affected by a pronounced homogeneity of the reservoir rock, as this will keep the gas from "breaking through" to the well via highly permeable channels.

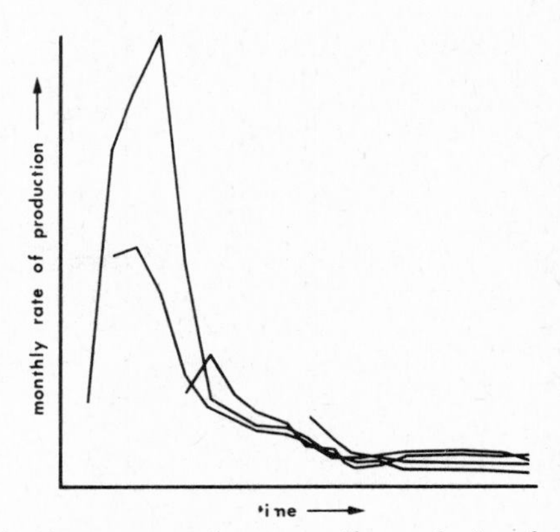

Fig. 83 Production curves of some wells producing from a relatively small sand lens closed all around in a European oil field. The sand lens carries no water. Production exclusively results from gas expansion. Peak production including its relatively steep drop is called "flush production", usually followed by a period of low "stabilized production"

Whenever the oil is saturated with gas, any excess of gas will lead to the formation of a genuine "primary gas cap". Whereas in the case of undersaturated oil the first energy phase will be governed by the expansion of oil + dissolved gas, a saturated oil (crowned by a primary gas cap) will in the first phase be governed by the expansion of the gas cap.

In this case, production characteristics are largely determined by the size of the gas cap or, in other words, by the ratio (frequently designated by the letter "m") between gas volume and oil volume (both volumes under reservoir conditions).

Where m is very small – in other words where the gas cap is of small size – the phenomena will be similar to gas drive fields. Where, on the other hand, the gas cap is of substantial size as compared with the oil zone, the following will occur: (a) reservoir pressure will decline more slowly, (b) as a result of this, gas liberation will be retarded, (c) oil saturation and, along with this, effective oil permeability will remain higher and (d) the oil's viscosity will remain lower. When, in addition to these phenomena, (e) the structural relief is pronounced and (f) the reservoir rock is of substantial thickness and when, to end up with, (g) production is not forced and permits the gravity forces to act – then recovery may rise to 60 % and more. The wells will flow much longer than those in dissolved gas drive fields and, over and above this, gas cap action may be supported and prolonged by gas injection.
The wells closest to the gas/oil contact will show a rapid increase of the gas/oil ratio and soon "go to gas". Where no water drive is available wells drilled at structurally low points will give the best results, in particular when gravity comes into action.

2.5.3 Water drive

When, subsequent to the start of production, edge water encroaches into the reservoir and replaces, entirely or in part, the volume withdrawn from the reservoir by production, this (recovery mechanism) is termed "edge water drive". As a rule the edge water which, in many cases, also underlies the oil, is exposed to a substantial hydrostatic pressure usually equalling the difference in height between the edge water level and the outcrop of the rock at the surface (refer to Fig. 70). The same applies to the "bottom water" underlying the oil in reservoirs of little dip and with a thick oil-bearing stratum.
For the flow of the edge water or, in other words, for its influx into the producing oil- or gas-field, two sources of energy may be responsible:

1. An outward source offered by the outcrop of the rock at the surface which causes the water to be replenished from the outside.
2. Inherent sources of energy in the water itself such as (a) the expansion of the water which always contains at least some dissolved gas, (b) the expansion of the liberated gas or of gas pockets and, finally, (c) the compressibility of the water-bearing layer itself or, in other words, a reduction of the formation's pore space as pressure declines.

An ideal example of such an inner source of energy is offered by the East Texas field. Although the Woodbine sand is exposed at the surface, pressure loss over such enormous distances has been computed to be so substantial that the sole force responsible for the edge water drive in this field is the expansion of the water and not,

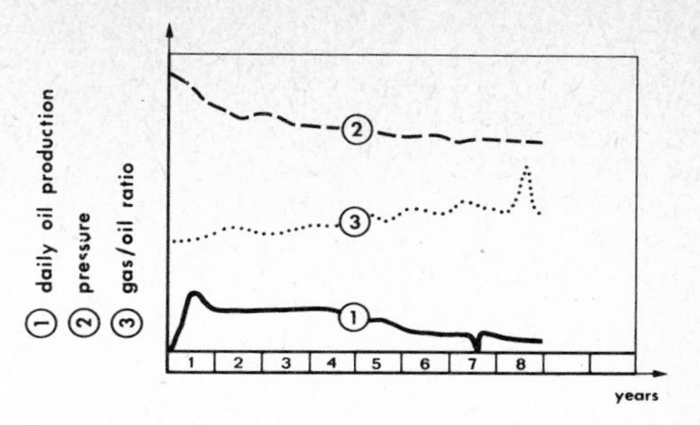

Fig. 84 In a Gas Cap Drive Reservoir the daily rate of production is slowly decreasing (depending on size of the gas cap); the reservoir pressure falls slowly and continuously; the GOR rises continuously in up-structure wells. Water production is negligible. Recovery factor 20–40 %. (After *Clark*[4])

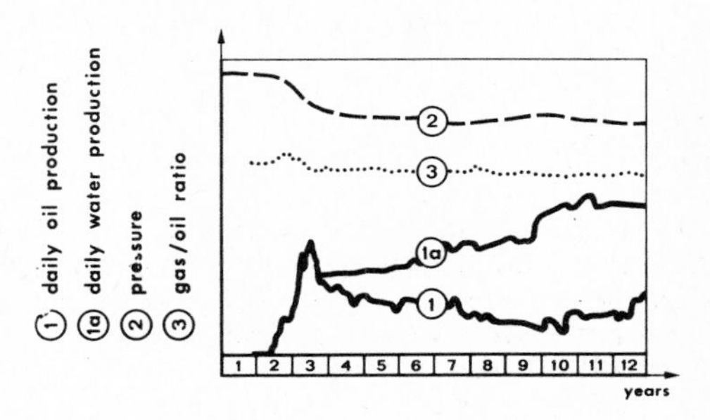

Fig. 85 In a Water Drive Reservoir the reservoir pressure remains high; GOR does not change as long as the bubble point pressure is not reached; water production starts at structurally low wells and increases steadily; recovery 35–75 %; rate of production influences pressure: at low production rates, reservoir pressure remains constant; at fast rates it is decreasing and, at reduced rates, the pressure will increase. (After *Clark*[4])

as one might be led to believe, the fresh supplies of water coming from the outcrop. Practically the same applies to most water drive fields, for, as a rule, the oil reservoir is relatively small as compared with the surrounding water bodies the extent of which is usually large.

Aquifers may be classified as follows:

1. The reservoir rock crops out at the surface ("open reservoir").

a) Replenishment of the water reservoir is unlimited or, at any rate, exceeds the volume withdrawn from the reservoir by production.

b) Replenishment of the aquifer from the surface is limited or the water level does not reach the surface (for instance in the case of the Karst water level of the "South German Jura").

2. The water reservoir is limited, usually because the reservoir rock becomes impermeable.
a) As compared with the volume of the oil in the reservoir the volume of the water is very large. Under these circumstances the edge water drive will be practically unlimited.
b) As compared with the volume of the oil the volume of the water is insignificant.

However desirable it may be to know as early possible whether a field may be produced on a natural water drive or not, conditions in practice usually render such predictions difficult, if not impossible. When in the course of constant production reservoir pressure remains constant, a natural water drive may be concluded to exist with certainty. The same applies when during a suspension of production, reservoir pressure is found to rise again. Under such conditions one will frequently succeed in keeping reservoir pressure constant by restricting the volume withdrawn from the reservoir (gas + oil + water) to the volume of the edge water flowing in. As a rule, the start of commercial production will involve a severe drop in pressure which, in the case of water drive fields, will then flatten out when the edge water has been set in motion. (Occasionally, the field engineer will be faced with the fact that very thick oil/water transition zones block the edge water for some time. Unfavourable saturation conditions may involve a relative permeability of so low a value that the edge water just does not begin to flow. In such cases, nothing will happen until the pressure gradient is large enough for this resistance to be overcome).
A natural water drive is the ideal source of energy for an oil reservoir. It follows that rational oil production requires this drive to be fully utilized and, in frequent cases, to be supported by an injection of water into the edge water zone. Under favourable conditions (homogeneous reservoir rock and favourable viscosity conditions, refer to Fig. 65) water drive fields obtain a recovery factor of up to 80 %.

2.5.4 Other forces and combinations of several forces

In reservoirs containing undersaturated crude, the first source of energy is the expansion of oil + gas in solution. Compared with the replacement of oil by gas or water, its significance, however, is restriced. Nevertheless, the first phase of production from undersaturated reservoirs is characterized by a rapid pressure decline.
The force of gravity is effective at any point in a reservoir. Even prior to the start of production it gives evidence of its action by forming gas caps which are underlain by zones of oil and water.

In the course of production a further separation into gas and oil may take place. The steeper the dip of the layers, the larger their thickness, and the slower the flow movements, the stronger will be the action of this force, although, as a matter of fact, capillary forces also play a role in this connection.
The gravity force is, roughly speaking, the density difference between oil and gas (or water and oil) acting against the retention by capillary forces.
Theoretically the force of gravity permits the recovery factor to reach high values (up to 80 %). In the Federal Republic of Germany some of the older fields, in particular salt plug flank fields, are produced by force of gravity. In such fields the wells located at the lowest points of the reservoir structure are the most prolific ones.
When liquid pressure in the pores declines the rock's compressibility may reduce the total pore space ("effective pore compressibility"). Although this "source of energy" is somewhat theoretical, material balance calculations at least have to take it into account (refer to P. 14).
In general, nature causes a combination of sources of energy to become active either simultaneously or consecutively. In this way, in an undersaturated reservoir, the first drive mechanism will be that of oil + dissolved gas expansion. Later on, this will change into a gas drive (when pressure drops below bubble point), into an edge water drive, a gas cap drive or even into a combination of all of these. If the water drive available is strong enough, a dissolved gas drive may be prevented from materializing by keeping reservoir pressure at a level exceeding bubble point pressure.

Bibliography

[1] *Amyx, J. W., D. M. Bass, R. L. Whiting:* Petroleum Reservoir Engineering. Mc Graw-Hill, New York 1960.

[2] *Baldizár, T.:* Geothermal Energy Production from Porous Sediments in Hungary. Geothermics, Vol. 2, Part. 1, 1970.

[3] *Becker, J.:* Untersuchungen zur Untertagegasspeicherung in einem wasserführenden Sandstein. Erdöl-Erdgas-Zschr. Sept. 1965.

[4] *Clark, Norman I.:* Elements of Petroleum Reservoirs. AIME, Dallas 1960.

[5] *Collins, R. E.:* Flow of Fluids through porous Materials. New York 1961.

[6] *Engelhardt, W. v.:* Die Bildung von Sedimenten und Sedimentsgesteinen. Schweizerbart, Stuttgart 1973.

[7] *Fatt, I., D. H. Davis:* Trans. AIME 1952.

[8] *Frick, Th. C.:* Petr. Prod. Handbook. Mc Graw-Hill Co., New York 1962.

[9] *Füchtbauer, H.:* Sediments and Sedimentary Rocks 1., Schweizerbart, Stuttgart 1974.

[10] *Gaida, K.-H., W. Rühl, W. Zimmerle:* Rasterelektronenmikroskopische Untersuchungen des Porenraumes von Sandsteinen. Erdöl-Erdgas-Zschr. Sept. 1973.

[11] *Hubbert, M. K.:* Bull A.A.P.G., August 1953.

[12] *Kappelmeyer, O., R. Haenel:* Geothermics with Special Reference to Application. Borntraeger, Berlin 1974.

[13] *Knaap, W. van der:* Der Einfluß einer Änderung des Porenvolumens in erdölhaltigem Gestein auf die Abschätzung des Ölvorrates. Erdöl und Kohle, 1960.

[14] *Knutson, C. F.:* AAPG, Sept. 1954.

[15] *Levorson, A. I.:* Geology of Petroleum. Freemann & Co., San Francisco 1967.

[16] *Meder, H.:* Über die Berechnung der Permeabilität von Sandsteinen aus Porosität und Korngrößenverteilung. Dissertation, Clausthal 1965.

[17] *Pirson, S. J.:* Oil Res. Eng., New York–London 1958.

[18] *Rieckmann, M.:* Erdöl und Kohle, Erdgas, Petrochemie, Juni 1963.

[19] *Rizzi, P.:* Hochdruckzonen-Früherkennung in Mitteleuropa. In: Erdöl-Erdgas-Zschr., Juli 1973.

[20] *Roll, A.:* Geologisches Jahrbuch, Reihe A, Heft 14, 1974.

[21] *Saidi, A. M.:* Petr. International, Febr. 1975.

[22] *Schmid, C.:* Erdöl und Kohle, 5, 1952.

[23] Verhandelingen van het Nederlands geologisch mijnbouwkundig Genootschap, Deel 28, 1973.

[24] *Wright, H. T., H. D. Woody:* AIME Symposium on Formation Evaluation, Oct. 1955.

3 Flow of Fluids through the Reservoir

When starting production of a well and reducing mud column pressure to a point below reservoir pressure, the pressure gradient between the reservoir and the borehole causes the content of the pores to flow. The direction of flow and the rate of flow resulting from a given pressure gradient depend on certain physical features of the flowing medium (viscosity, phase behaviour, etc.) on the one hand, and on the nature of the reservoir rock (permeability, pore geometry, etc.) on the other.
Flow phenomena in reservoirs may vary between very simple processes (such as the linear flow of a non-compressible liquid through homogeneous rock) and combinations of such complexity that their evaluation is possible only with the help of complicated computer programs or reservoir models. This is easy to understand when we realize that these phenomena may be (and in the reservoir they usually are) complicated by the following variations:

1. Direction of flow. This may be
a) linear: the medium moves in linear flow in a single direction through a constant cross-section;
b) radial: the medium converges radially to a given point; the cross-section available for this kind of flow is obviously reduced steadily as it approaches the centre point;
c) spherical: while, in radial flow, the flow lines converge at the same level, spherical flow makes them converge in three dimensions, for instance on a single perforation hole in a well.

The pore structure of oil and natural gas reservoirs usually leads to directions of flow of a far more complicated nature. However, the above generalizations will do for mathematical derivations.

2. Nature of the flowing medium which may be
a) non-compressible;
b) compressible.

3. Flowing phase. The medium flowing may be
a) a single-phase medium (gas or oil or water) = homogeneous flow,
b) a two-phase medium (gas + liquid) = heterogeneous flow. In this case the fluid may be either oil or water only or oil + water.
c) the same applies accordingly when gas + oil + water flow.

4. Flow dynamics
a) Steady-state flow: if at a given point pressure remains constant flow has become steady (as quantity and velocity of flow remain constant).

If, on the other hand, a pressure drop occurs at this point and if this pressure drop is, per unit of time, slight (and constant) the flow materializing under such circumstances is called “semi-steady-state flow". This type of flow is generally treated as a variant of steady-state flow.

b) Unsteady-state flow: A sudden change of the pressure gradient causes a breakdown of "stabilized" conditions prevailing up to this change. Quantity and velocity of flow change continuously. Until conditions stabilize again this type of flow is called transient-state flow.

5. The reservoir (including the aquifer) may be either "infinitely large" or "finite", i.e. "closed". This simple alternative may be complicated by innumerable variants referring to the geometrical shape of the reservoir.

6. In addition to the above mentioned phenomena the rock itself may be uniformly permeable (isopermeable) or its permeability may vary strongly (heteropermeable). In the latter case the changes may occur either in a single direction (for instance parallel to stratification or at right angles to it) and are then called *isotropic;* or in all possible directions, when they are called *anisotropic.*
Factors influencing flow behaviour

homogeneous heterogeneous	flow
linear radial spherical	direction
isopermeable heteropermeable isotropic anisotropic	rock
steady-state transient-state unsteady-state	quantity/pressure/ velocity
compressible non-compressible	flowing medium

The above table illustrates how many factors are to be considered for an accurate appraisal of flow phenomena. The basis however, is always Darcy's Law in conjunction with the saturation conditions prevailing in the pores. For our purposes it will do to mention a single formula which has a direct impact on many problems confronting the production geologist in day-to-day practice. It describes the radial, homogeneous, steady-state flow of a non-compressible liquid:

$$q_{sc} = \frac{k \cdot h \cdot (p_e - p_w)}{\mu \cdot B \cdot \ln (r_e/r_w)}$$

in which:

q_{sc} = rate in cm^3/sec under standard conditions
k = permeability in md
h = thickness in m
p_e = pressure at the limit of the well's drainage area in bar
p_w = pressure in the well in bar
μ = viscosity in cP
B = formation volume factor
r_e = radius of well drainage area
r_w = radius of well

r_e may either be the radius of a circular reservoir (insofar as the pressure gradient around the well in the centre of the circle is propagated right to the limit of the reservoir); or it may be the radius of the well's drainage area (when under strong water drive the pressure gradient around the well just extends to a certain distance from the well); or again it may be the radius of the well's drainage area when the well is surrounded by other wells. As usually more than one value of this equation is known or may be measured, it will be clear that this formula generally permits the calculation of several important values such as delivery (well capacity), rock permeability, well drainage area, or the degree to which the well is affected by other wells, etc.

3.1 One phase flow

The classical example of one-phase flow is the deduction of Darcy's Law. This law determines the permeability of reservoir rocks and is treated on page 19. It will be readily seen that this experiment concerns a linear, steady-state flow of a non-compressible liquid through an isopermeable rock. When considering the phenomenon in question, the following factors should not be neglected, however:

a) Reynold's number
Darcy's Law is restricted to laminar flow, a flow governed by internal friction (viscosity) and not subject to forces of inertia (which cause turbulence phenomena). Whether flow is laminar or turbulent for a given volume V depends on a non-dimensional number. It is called Reynold's number and is the ratio of inertia to internal friction. The Reynold's number for sandstones, for instance, is:

$$N_{Re} = \frac{V \cdot p \cdot d_p}{\mu}$$

in which:

V = the flowing volume
μ = the viscosity of the flowing medium
$\overline{d}_p$ = average grain diameter of the sandstone

Transition from laminar to turbulent flow is gradual and ranges between $N_{Re} = 1$ and $N_{Re} = 10$. However in view of the extremely complicated pore geometry the forces of inertia are of increasing importance. This is why a pseudo Reynold's number N_{Rp} has been introduced. The latter emphasizes the cross-sectional area F, while the Reynold's number N_{Re} attaches more importance to the length l of the capillary tubes.

b) Knudsen flow and Klinkenberg effect
Permeability is a rock-specific constant or, in other words, a coefficient independent of the pore saturation. It follows that it should be possible to describe the flow of liquids or gases through a uniform permeability. This would then apply to flow phenomena of liquids which are inert to the rock matrix; also to gases under high pressure.
While, as a rule, liquids will "adhere" to the walls of the pores with their outermost layer, gases slip along these walls and so possess more flow velocity ("slipping constant" c) In addition to this, the following should be considered, as well: as a rule the molecules' proper motion (Brownian motion, mean free molecular path λ) is smaller than the pore radius r. In frequent cases, however, $\lambda > r$. Under these conditions the molecules "hit" the pore walls, they diffuse in the direction of the smallest pressure (Knudsen diffusion). This applies in particular to gases under low pressure and to very small pore radii. (Consequently under reservoir pressure, the diffusion flow should have no importance).
It follows that the permeabilities for air measured in the laboratory at an average gas (air) pressure Δp (Fig. 24) of 1–2 bar, are greater than the permeabilities determined with inert liquids. (In other words: gases under low pressures deviate somewhat from Darcy's Law). This penomenon common to porous media is called the "Klinkenberg effect" and is described in Appendix H.

c) Special conditions occur when electrolyte solutions flow through a rock containing clays and, in particular, montmorillonites. This is another case of deviation from Darcy's Law. In general, solutions with a small salt content (involving a low degree of electrolytic dissociation and a high pH) reduce the permeability, while solutions with a high salt content increase it. As yet, the individual causes of the phenomenon are little known. There is no doubt that electrokinetic effects materialize and that these may cause the clays to swell and to reduce permeability. It is also quite feasible to presume the generation of a current potential producing an electro-osmotic motion which, in counteraction to the direction of flow, reduces flow rate (also refer to Appendix I).

3.2 Multiple phase Parallel Flow

As a rule in a reservoir two or three media flow together: gas, oil and water. Their respective shares in the pore volume are symbolized by S_g, S_o and S_w (also refer to page 32).
It is not difficult to imagine that permeability for a given medium – let us say water – is affected by the presence of another medium – let us say oil – in the pores, and that under these conditions it is lower for the flow of the first medium than it would have been if the pore space were filled with water alone. Even if water is the sole medium to flow, the flow channels for its passage are so to say narrowed by the presence of the other medium. This "reduced" permeability is either called "effective permeability" for the flowing medium (if expressed in Darcy's) or "relative permeability" (k_{rw}, k_{ro}, k_{rg}) when written as a percentage or a decimal fraction of "absolute permeability" (or rock-permeability). For the purpose of measuring relative permeability, a constant quantity of oil + water is pumped through a rock sample. As, in this way, oil and water flow alongside each other this phenomenon may be called "parallel flow".
Although in a reservoir such parallel flow will occur for certain periods of time and in certain restricted sections, the oil, for instance in a water-drive field, is in the end displaced by the encroaching edge water. The same is the case in water flooding operations. There

absolute permeability k

effective permeability $k_g \quad k_o \quad k_w$

$$\text{relative permeability} = \frac{\text{effective permeab.}}{\text{absolute permeab.}}$$

$$= k_{rg} \quad k_{ro} \quad k_{rw}$$

$$= \frac{k_g}{k} \quad \frac{k_o}{k} \quad \frac{k_w}{k}$$

$$k_g + k_o + k_w < k$$

$$k_{rg} + k_{ro} + k_{rw} < 1.0 \quad (< 100\%)$$

$$S_g + S_o + S_w = 100\% = V_p$$

Fig. 86

are several displacement theories permitting an arithmetical coverage of the processes brought about by such operations.
After all, even the material balance, a volumetric balance of reservoir contents at various times (or, in other words, under various conditions of pressure and saturation) is a kind of displacement calculation. The volume withdrawn from the reservoir by production is replaced somehow, and no vacuum is produced in the pores. This displacing medium need not necessarily be water. It may be expanding oil or gas or even gas liberated from oil.

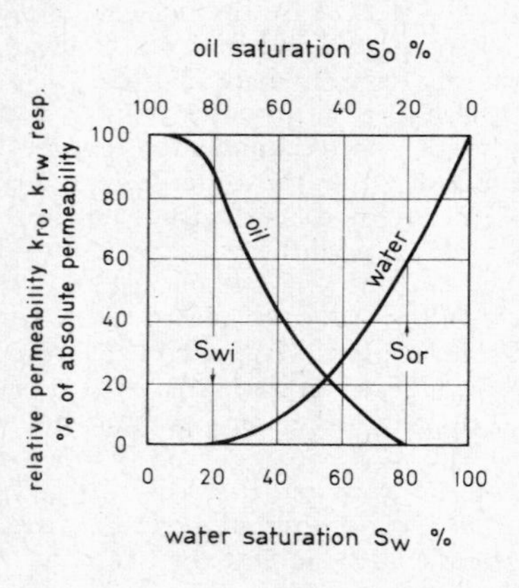

Fig. 87 Relative Permeability for water (k_{rw}) and oil (k_{ro}) vs. saturation (S_w and S_o resp.)

This diagram offers direct readings of relative permeabilities for oil and water at various degrees of saturation, e.g.:

		1	2	3	4
S_w	%	20	40	55	80
S_o	%	80	60	45	20
k_{rw}	%	0	8	23	60
k_{ro}	%	75	45	23	0

Up to a water saturation of 20 % no water flows (= S_{wi}). No oil will flow when the oil saturation falls below 25 % (= S_{or}).

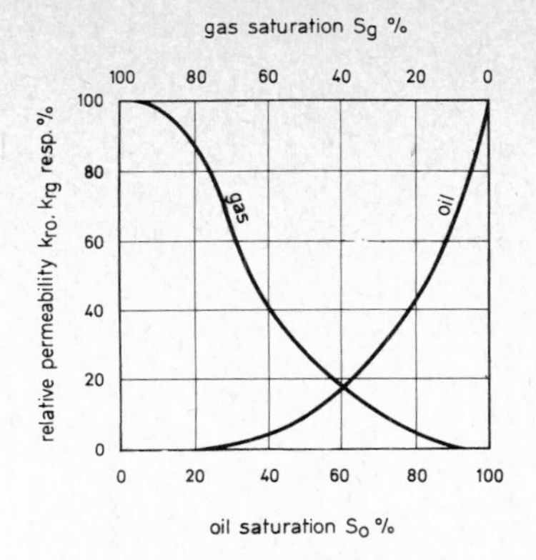

Fig. 88 shows relative permeability for oil and gas. It indicates:
up to $S_o = 20\,\%$ the sole medium flowing in the rock is gas
from $S_o = 20\,\%$ to $90\,\%$ both oil and gas are flowing in the rock
as from $S_o = 90\,\%$ the sole medium flowing in the rock is oil

3.2.1 Relative (and effective) Permeability

On page 19 we have seen that permeability, like porosity, is a specific feature of the reservoir rock. If, for instance, we have a porosity of 25 % and a permeability of 500 md, these values outline (together with the P_c graph) the reservoir rock's properties essential to the petroleum engineer.
Beyond this, we have seen that (when observing certain precautions) we always arrive at the same permeability value, no matter whether a gas or liquid is used for its measurement. This is why occasionally this rock parameter has been designated as the "absolute permeability k". A prerequisite for these measurements, however, is that only *one* medium flows through the rock or, in other words, that the pore space is completely saturated with *one* medium:

S_o or S_w or S_g must equal 100 %.

Where this is not the case, that is, where two or three different media fill the pores, we no longer obtain the absolute permeability, but a different (and somewhat lower) value designated as the "effective permeability" (when expressed in Darcys) for a certain flowing medium.

The ratio described by the formula $\dfrac{\text{effective permeability}}{\text{absolute permeability}}$

is called "relative permeability". As a rule it is expressed by a decimal fraction, often also by a percentage of absolute permeability. This ratio has to be designated with the symbol for gas (g), oil (o) or water (w) and so it becomes k_{rg}, k_{ro} or k_{rw}. Fig. 86 summarizes the symbols and certain relations.

In Fig. 87 relative permeabilities for oil and water are plotted in a diagram indicating water saturation and the corresponding oil saturation in %. The immobile "irreducible water saturation" is designated by the symbol S_{wi} (or S_{wc} = connate water). It increases as the reservoir rock's permeability decreases. The irreducible oil saturation (or, in other words, the oil no longer recoverable by standard production from the reservoir) is designated as "residual-oil saturation" = S_{or}.

It should be stated that the sum of the effective permeabilities is always lower that the absolute permeability. In addition to this,

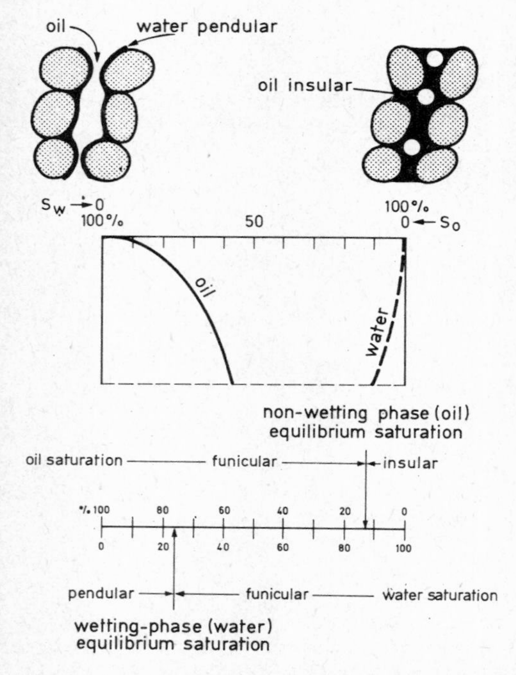

Fig. 89 An interesting phenomenon is that the curve of relative permeability for oil (k_{ro}), when oil saturation (S_o) is high, differs from that of relative permeability for water (k_{rw}), when water saturation (S_w) is high. This is explained by the different distribution of wetting phase (water) and non-wetting phase (oil). The terms pendular, funicular, insular saturation are taken from S. J. Pirson: Oil Reservoir Engineering (loc. cit.)

the two curves do not have the same (mirror-inverted) shape. The reason is that one of the liquids is wetting while the other is not. (As the non-wetting phase is in the larger pores, its effective permeability is always greater than that of the wetting medium).
Another look at relative permeabilities for oil and water as represented in Fig. 89 shows that the curves of relative permeabilities for oil (at high oil saturations) and for water (at high water saturations) have quite different shapes. This is explained by a different distribution of the flowing media in the pores (refer also to Fig. 47).
Although the relative permeability curves for the various reservoir rocks resemble each other in principle, there are certain differences deriving from the individual rock's "pore geometry". This is demonstrated in Fig. 90.

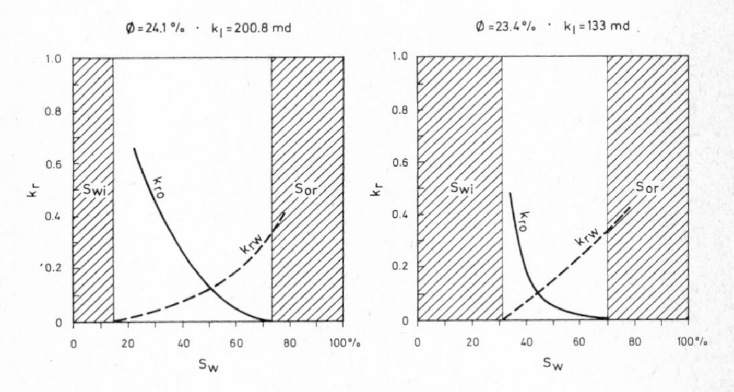

Fig. 90 The rock shown in this figure on the left is a Lower Cretaceous sandstone with a porosity of roughly 24 % and a permeability of 2008 md. In such permeable rocks the connate water content (S_{wi}) is small, in our case just 15 %. Residual oil saturation (S_{or}) is 27 %. At a saturation of oil and water (S_o and S_w resp.) of 50 % for either medium, the relative permeability for water (k_{rw}) is 0.125. On the right the same sandstone (ϕ 23.4 %) is investigated in a sample of far lower permeability (just 133 md). In this case, the connate water content S_{wi} amounts to 31 % while residual oil saturation S_{or} equals 30 % and $k_{ro} = k_{rw}$ occurs at a point where water saturation $S_w = 44$ % and oil saturation $S_o = 56$ %

Summarizing: relative permeability for flowing media depends in the first place on the saturation. Apart from saturation, the shape of the k_r curves is also (in a subordinate degree) determined by the pore geometry and by the surface tension of the media filling the pores.

Fig. 91 shows a well-known method of measuring relative permeability.

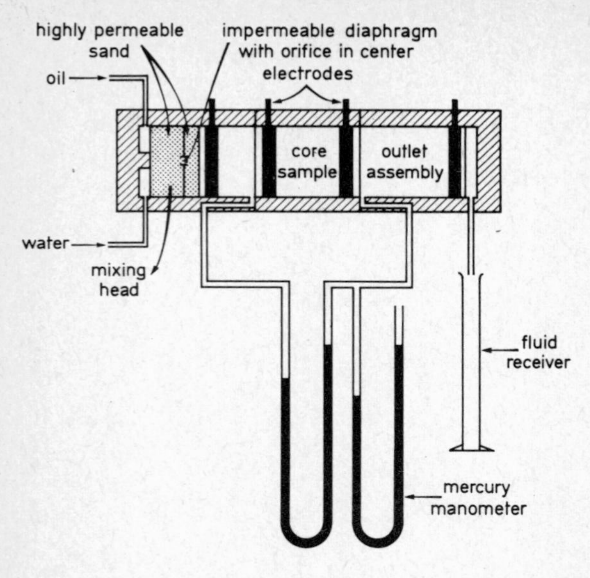

Fig. 91 Apparatus for measuring the relative permeability of simultaneous flow of two fluids (after *Pirson*[11]). Relative permeability measurements on rock samples are faced with manifold difficulties such as:
the "*end effect*": the capillary effect is interrupted at the end of the rock sample. Perhaps, this is one of the reasons why, in some cases, different values are observed at different rates of flow;
the "*hysteresis effect*": a discrepancy of values measured for, let us say, a water saturation of 50 % depending on whether this percentage is reached by a "fill-up" from 0 to 50 % or by a reduction from 100 to 50 %

3.2.2 Mobility and f-Functions

According to Darcy's Law not only the relative permeability, but also the viscosity of flowing media must be taken into account. The ratio of relative permeability to viscosity is called the "mobility factor λ", in other words:

$$\text{mobility factor of water} = \lambda_w = \frac{k_{rw}}{\mu_w}$$

$$\text{mobility factor of oil} = \lambda_o = \frac{k_{ro}}{\mu_o}$$

When oil and water flow together, the flow rate of each medium (at a given saturation) depends on the "permeability-viscosity ratio" which is called "mobility ratio M". This ratio is expressed by the following formula:

$$M = \frac{k_{rw} \mu_o}{k_{ro} \mu_w} = \frac{\lambda_w}{\lambda_o}$$

To give an example with the following data

$S_w = 70\,\%$
$S_o = 30\,\%$
$k_{rw} = 0.225$
$k_{ro} = 0.045$
$\mu_w = 5$ cP
$\mu_o = 1$ cP

this means:

$$M = \frac{0.255 \cdot 5.0}{0.045 \cdot 1.0} = 15$$

or, in other words, water flow is 15 times greater than oil flow. As relative permeability varies along with saturation, and as the latter changes permanently, the mobility ratio also changes permanently. M is the factor governing any replacement of oil by water,

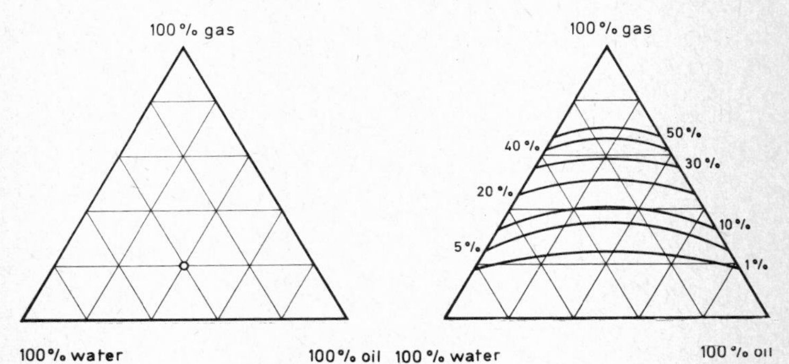

Fig. 92 Relative permeability can be represented if oil, water and gas are present in the pores of a reservoir rock with varying saturation. For this purpose saturation values are shown in the shape of a triangle (left-hand figure). The right-hand figure represents relative permeability for gas at different saturations. What should be remarked is that effective permeability for gas is just about 50 % of absolute permeability even at the point where 70 % of the pores are filled with gas. (After *Leverett* et al.[6])

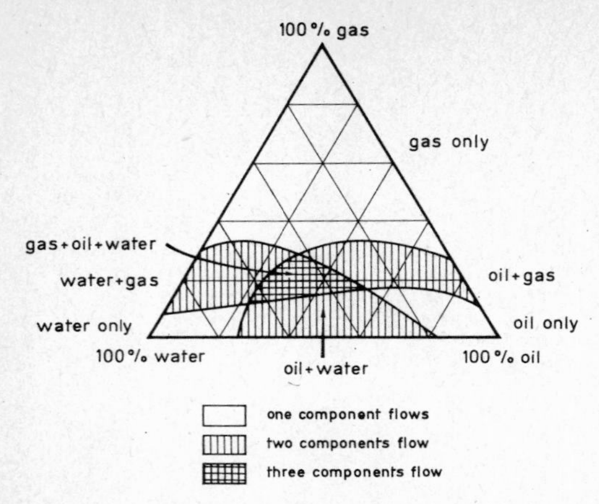

Fig. 93 shows that there are certain values of saturation permitting just one component to flow while the others are stationary. Other values permit two and still others three components to flow. (After *Leverett* et al.[6])

whether with natural water drive or water flooding. Efficiency is best under the conditions of so-called "stable replacement" or, in other words, when M = 1. The nearer the value for M is to 1, the better are the chances of a flooding project.
By introducing the mobility ratio it is possible to calculate to what extent water (f_w) or oil (f_o) contribute to total flow (q_{tot}). Obviously, the total quantity flowing (q_{tot}) is composed of the quantity of water (q_w) and the quantity of oil (q_o), so that

$$f_w = \frac{q_w}{q_{tot}} = \frac{q_w}{(q_w + q_o)}$$

An introduction of viscosities and relative permeabilities into this equation results in

$$f_w = \frac{1}{1 + \frac{\mu w}{\mu_o} * \frac{k_{ro}}{k_{rw}}}$$

$$f_o = \frac{1}{1 + \frac{\mu_o}{\mu w} * \frac{k_{rw}}{k_{ro}}}$$

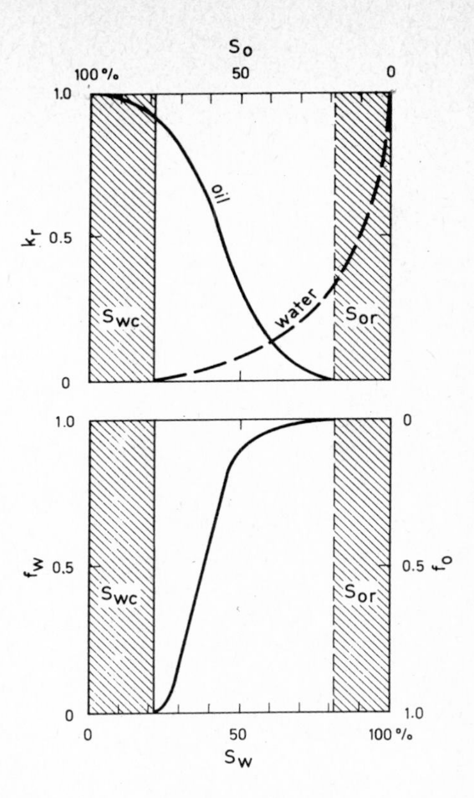

Fig. 94 shows the relation between relative permeability and the components of oil and water flowing under various conditions of saturation. It shows the connate water contents (irreducible water saturation S_{wi}) to be 22 % in this example while the residual oil saturation (S_{or}) amounts to 18 %. Relative permeabilities for oil and water (k_{ro} and k_{rw} respectively) are represented by curves in the usual way. Water does not begin to flow until water saturation exceeds 22 % while oil ceases to flow at a saturation of 18 %.

Under the same conditions of saturation, the proportions of water f_w and oil f_o in total flow are represented in the lower curve. It shows that as water saturation increases, flow of water goes up rapidly, and when water saturation amounts to just 39 % while oil saturation is still 61 %, flow is already composed of 50 % of water and oil each. Ingress of water then increases very rapidly, and at a water saturation of 50 %, flow already contains 88 % of water

Viscosity Ratio (water : oil) $\frac{\mu_w}{\mu_o}$

(oil : water) $\frac{\mu_o}{\mu_w}$

$$\text{Mobility Factor } \lambda = \frac{\text{relative permeability}}{\text{viscosity}} = \frac{k_r}{\mu}$$

$$\lambda_w = \frac{k_{rw}}{\mu_w}$$

$$\lambda_o = \frac{k_{ro}}{\mu_o}$$

Mobility Ratio M
usually displacing fluid : displaced fluid
e. g. water : oil

$$M \text{ (water : oil) } \frac{\lambda_w}{\lambda_o} = \frac{\frac{k_{rw}}{\mu_w}}{\frac{k_{ro}}{\mu_o}} = \frac{k_{rw}}{k_{ro}} \frac{\mu_o}{\mu_w}$$

$$M \text{ (oil : water) } \frac{\lambda_o}{\lambda_w} = \frac{\frac{k_{ro}}{\mu_o}}{\frac{k_{ro}}{\mu_w}} = \frac{k_{ro}}{k_{rw}} \frac{\mu_w}{\mu_o}$$

f – Functions

$$f_w = \frac{1}{1 + \frac{k_{ro}}{k_{rw}} \frac{\mu_w}{\mu_o}} = \frac{1}{1 + M \text{ (oil : water)}} = \frac{1}{1 + \frac{1}{M \text{ (water : oil)}}}$$

$$f_o = \frac{1}{1 + \frac{k_{rw}}{k_{ro}} \frac{\mu_o}{\mu_w}} = \frac{1}{1 + M \text{ (water : oil)}}$$

Fig. 95

The result, the value of f, may be expressed either as an percentage (for instance 95 % "water cut") or as a decimal fraction (f_w = 0.95) (see Fig. 96).

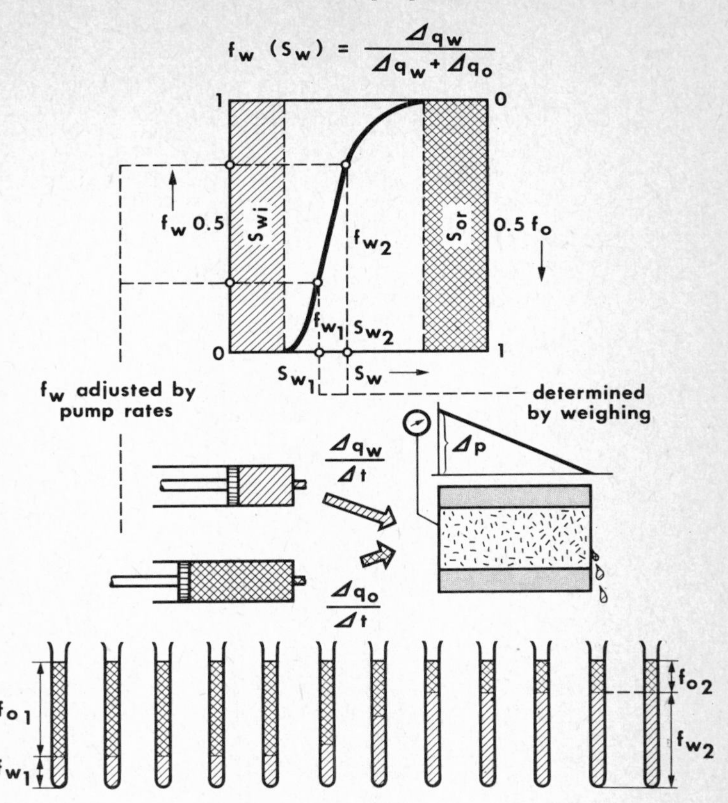

Fig. 96 Determination of Function $f_w = f_w\ (S_w)$. The simple-hatched areas represent water, whereas the cross-hatched areas stand for oil. (By courtesy of Dr. *Miessner*, Brigitta/Elwerath, Hannover)

From the above formula the importance of the viscosity ratio between oil and water $\dfrac{\mu_o}{\mu_w}$ may be readily inferred. As this viscosity ratio largely depends on temperature, it is obvious that this is key to all thermal processes of oil recovery (refer to Page 104).

Let us summarize on the basis of a practical example:

Fig. 97 shows a steeply dipping oil-bearing rock with a "closure" of 60 m (i.e. an oil-bearing part with a difference of elevation of 60 m between the edge water line and the seal provided by an impermeable clay). Opposite to it, the figure shows the capillary-pressure curve of the reservoir rock (which we assume to be homo-

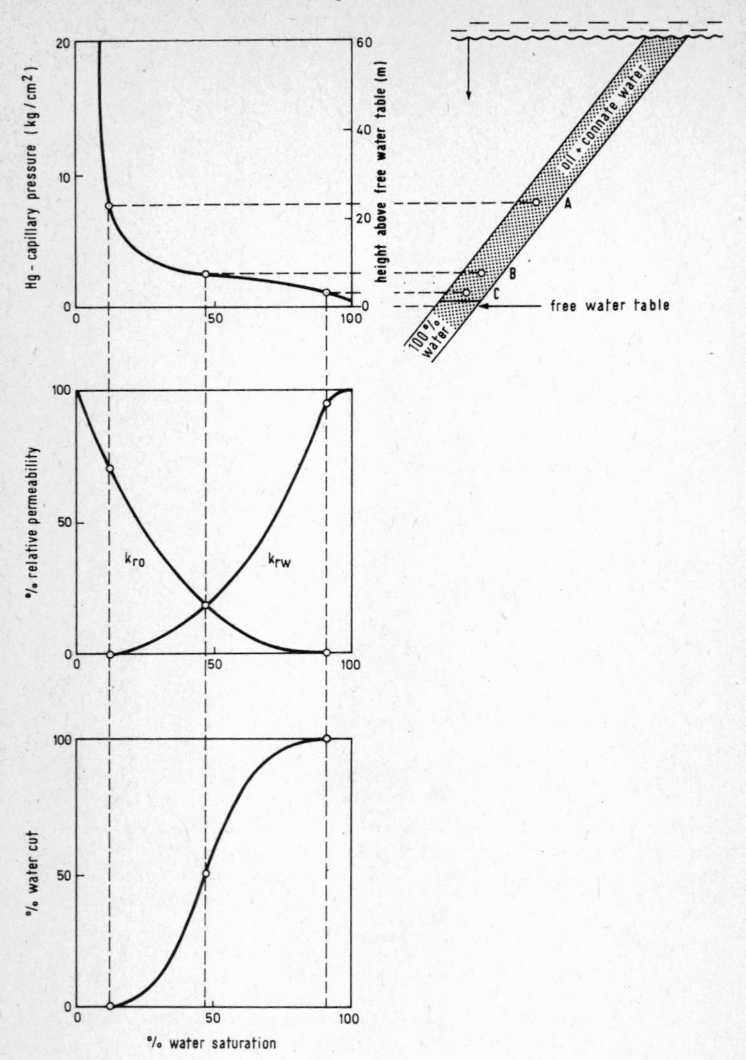

Fig. 97 (see page 91)

geneous so that each sample of this rock would show the same P_c curve). As set out on p. 45, this capillary-pressure curve permits us to compute the elevation in meters over free edge water level

or to compute the position of this level if it is not yet known from the results of previous wells. Values of saturation prevailing at any point of the reservoir may be determined with the help of the P_c curve (at point A, for instance, 10 % water saturation and consequently 90 % oil saturation).
Relative permeability for oil and water, k_{ro} and k_{rw}, may be determined by lab tests and may also be correlated with the saturations

S_o and S_w.

An application of the formula to the calculation of the water component f_w on p. 88 permits us to plot the percentage of this component in total production.
A well striking the reservoir at point A, for instance, will have no influx of water although water saturation S_w is 10 % at this point. At point B, on the other hand, water saturation is 45 % and $k_{ro} = k_{rw}$ at a viscosity ratio $\mu_w/\mu_o = 1$. At this point 50 % of production will be water. At a point like C, finally, we produce nothing but pure water, although 10 % of the pore space is still filled with oil.

3.3 Displacement of oil (and gas) by water

In certain parts of the reservoir and over limited periods of time, a "parallel flow" of gas, oil and water alongside each other occurs. However, when considering the reservoir as a whole, the oil is volumetrically forced to the well by gas and water and so is actually displaced from the reservoir.
The volume withdrawn from the reservoir by production must be replaced somehow, as no vacuum can occur in the pore space. This "substitute volume" may either be available in the reservoir itself (such as expansion of oil and dissolved gas, liberation und subsequent expansion of gas, reduction of the pore space due to the compressibility of the reservoir rock) or it may penetrate into it from the outside (for instance from the gas cap, by encroaching edge water or by flooding). In the latter case, the process might remind one of a piston forcing a liquid through a pipe.
However, the media encroaching from the outside never displace the oil flowing to the wells as completely as would do a piston sweeping the inside of a pipe. The advancing medium follows channels it prefers for geometrical or physical reasons and, at first, in numerous pores, leaves oil which is then gradually removed by fresh supplies of water or gas. As, in the course of this process, there is a steady increase of the volume of the displacing medium, while the proportion of oil in overall production decreases correspondingly. This process – formally described by relative permeability as a function of saturation – increasingly impairs the

efficiency of the displacement mechanism (down to the economic limit). The principle of this displacement process in a pipe and in a porous rock is schematically described in Appendix K.
In their role as displacement media water and gas differ in several respects. When oil is present gas will not wet the reservoir rock while water is, as a rule, the wetting medium. Furthermore the (generally large) difference in viscosity between water and gas affects the displacement mechanism. Much more than the more viscous water, gas will tend to flow mainly by "paved" tracks and bypass the oil. (The viscosity ratio between gas and oil lies, as a rule, between 100 and 10,000; while the ratio $\frac{\mu_w}{\mu_o}$ is generally between 1 and 100.)

3.3.1 Displacement Theories

To describe the displacement process, several theories have been elaborated, in particular with respect to the displacement of oil by water. Such displacement theories permit a computation of future production performance as well as of the expected recovery. The various methods use parameters which, in themselves, are more or less easy to determine, and all of them offer simplified approximate solutions. They describe the displacement process by mathematical relations part of which were developed from model concepts. The following table covers the essential characteristics of some displacement theories (table 1).
Buckley & Leverett[6] developed a displacement theory based on relative permeability. They start out from a water drive reservoir in which edge water advances in frontal fashion (linear case) and in which total flow rate (oil + water) is constant in every cross section. Their basic assumption is the f-functions (refer to page 88). Capillary forces and gravity are not considered.

$$f_w = \frac{1}{1 + \frac{k_{ro}\mu_w}{k_{rw}\mu_o}} \quad \text{or}$$

$$f_o = \frac{1}{1 + \frac{k_{rw}\mu_o}{k_{ro}\mu_w}}$$

$$= \frac{1}{1 + M}$$

$\frac{\mu_w}{\mu_o}$ is constant (for a given reservoir pressure)

Table 1

Theory advanced by	*Buckley-Leverett*	*Dietz*	*v. Meurs v. d. Poel*	*Marsal*
Name	Frontal Advance Rate	Water-Tongue	Viscous Fingering	–
Field of application	Linear advance	linear	Linear	Linear advance and advance via well pattern
Method	Numerical	Analytical	Analytical	Extrapolating numerical
Basis	Relative permeability	Water tongue	Water finger	Production behaviour of wells in low parts of reservoir structure
Considers				
viscosity	via relative permeability (viscosity, mobility, permeability)	+	+	considered via production behaviour (viscosity to well net)
mobility		+	–	
permeability		+	+	
Stratification	–	–	–	
gravity	hardly	hardly	in direction of flow	
residual water	via relative permeability	+	+	
capillary forces	–	–	–	
well net	–	–	–	
Essential components of formula	relative permeability	mobility	viscosity, residual oil saturation, minimum water saturation	production behaviour oil-bearing pore space
Remarks	1. Applies to homogeneous reservoirs, only 2. Results uncertain, as relative permeability is hard to measure	1. homogeneous reservoir 2. Water breakthrough indicated too early	disregards connate water	for constant pay only

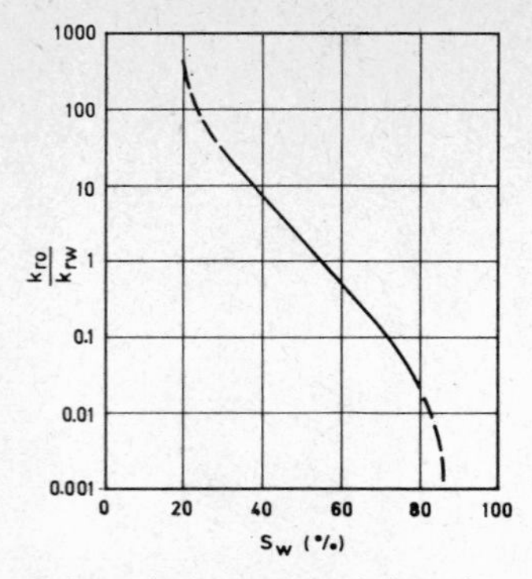

Fig. 98 The ratio of the relative permeability for oil to that of water (k_{ro}/k_{rw}) depends on the water saturation. The fraction of water flow (f_w) is, therefore, a function of S_w

$\frac{k_{ro}}{k_{rw}}$ depends on water saturation S_w (see Fig. 98)

f_w, therefore is a function of S_w

Via the continuity equation, *Buckley & Leverett* succeeded in deriving a formula permitting the calculation of the distribution of water saturation in a reservoir at a given time. This formula reads:

$$x = \frac{q \cdot t}{A \cdot \phi} \left(\frac{df_w}{dS_w}\right)$$

In this equation, x is the distance of a plane of certain saturation from the "edge water line". S_w advances in the reservoir in a time t when the overall flow rate (oil + water) equals q. In other words: when the "edge water line" advances from left to right then a "slice" of a certain oil/water saturation will advance correspondingly from left to right, while preserving the constant saturation conditions particular to this individual slice. Differentiation then leads to the equation

$$\frac{df_w}{dS_w} = f(S_w)$$

It follows that the above equation is capable of solution, as Fig. 99 illustrates.

The method proposed by *Buckley & Leverett* can be simplified substantially[15]. In the first place, water saturation at the front of the flood can be found by applying a tangent to the curve $f_w = f(S_w)$ which passes through the point of origin ($f_w = 0\,\%$, $S_w =$

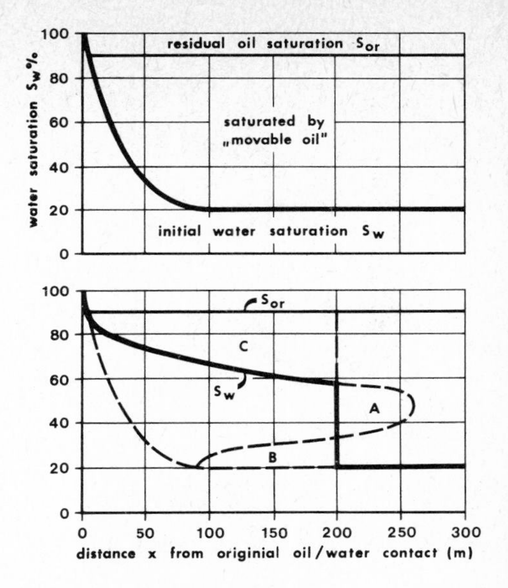

Fig. 99 Displacement theory by *Buckley & Leverett* (after *Buckley* et al.[1]). The upper figure shows the initial saturation: S_w is 100 % at the original oil/water contact and decreases to 20 % at a distance of 100 m from the OWC. The residual oil saturation S_{or} is 10 %. The lower figure illustrates the water saturation after a certain period of time (in *Buckley's* example after one year of production. The flooding front has not yet reached the structurally lowermost row of wells; therefore water break-through [BT] did not yet take place). *Buckley & Leverett* have artificially straightened the water front (by assuming area A = B). Area C represents the oil volume which could be additionally recovered (in the case of infinite water flow). By way of integration or planimetry one obtaines the quantity of recoverable oil (average water saturation minus initial water saturation) compared to the cumulative flow (q · t)

20 %). The point of tangency then gives the water saturation at the "front of the flood". In our example S_{wf} = 60 % (refer to Fig. 99 and Fig. 100).

The *Buckley-Leverett* method, however, is also applicable to the displacement of oil by gas. Due to the high oil/gas viscosity ratio $\frac{\mu_o}{\mu_g}$ as well as to the high relative permeability ratio between gas and oil $\frac{k_{rg}}{k_{ro}}$ when gas saturation is low, the degree of oil recovery is then,

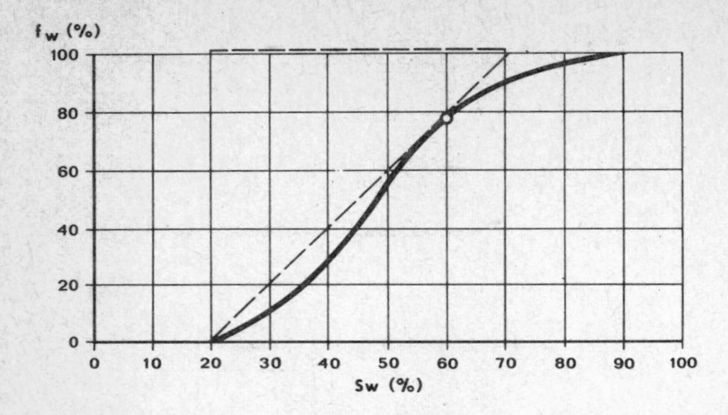

Fig. 100 The curve of f_w vs. S_w is S-shaped when capillary pressure is neglected. The place where the saturation changes abruptly is called the "front" and $S_{w\ front}$ is found by drawing a tangent from S_{wi} to the S-shaped curve. $S_{w\ front}$ advances with the same values and ahead of it the saturation is still original. A prolongation of the tangent shows the amount of recoverable oil: 70–20 = 50 %. (After *Welge* et al.[15])

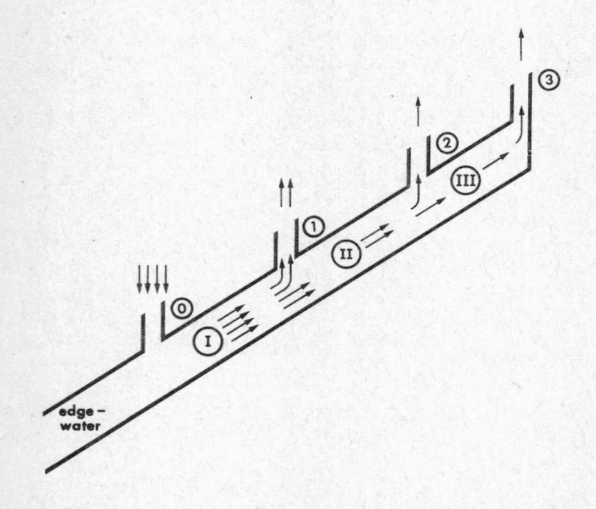

Fig. 101 Schematic representation of the "Marsal Method": let us assume a system of one injection well (0) and three producing wells (No. 1, 2 and 3). The number of arrows is proportional to the rates of flow (oil + water) in the different parts (I, II, and III) of the system. (By courtesy of Dr. *D. Marsal*, Brigitta/Elwerath, Hannover. See also: *Marsal*[7])

as a rule, much lower than in the case of water drive insofar as the gas drive is not supported by gravity drive.

Marsal's extrapolation method[7] for establishing oil recovery is an extension of the *Buckley-Leverett* theory. It assumes a linear displacement process taking into account the mutual interaction of a row of producing wells. The method is easily understood with the help of Figure 101:

Three wells numbered 1, 2 and 3 produce oil and water by means of an edge water drive or as a consequence of the injection of water. It is assumed that the oil + water produced is completely displaced by the water invading the system, and that there is no loss of the invading water into other parts of the reservoir. As a consequence, in the region I beneath the producing well 1 the flow rate of oil + water is equal to the total production rate of all three wells 1, 2 and 3.

At well 1, the flow stream divides: part is removed by production, the rest flows up-structure. Hence, in region II between well 1 and well 2, the rate of flow of oil + water is equal to the rate of invasion minus the rate of production of fluids at well 1. Finally, the rate of flow in the last part III of the system is equal to the rate of production of well 3.

So, knowing or assuming the production rates of oil or oil + water during a certain increment of time, we can calculate the advance of a certain f_w-value during the time step Δt by means of the *Buckley-Leverett*-equation, using the correct df/ds-value and the flow rate q of the region where S_w is located. The relationship between fractional water flow f_w and df/ds may be obtained by applying the whole procedure in reverse order to a system with known production rates and water cuts, extrapolating the f_w-df/ds-curve obtained. The method has been applied successfully for years in two German oil fields involving hundreds of wells. Further, it has stood the test of a large-scale model at the Shell Rijswijk labs in the Netherlands[4].

For his theory, *Dietz*[2] assumes that the water penetrates into the oil reservoir in the shape of a water tongue. When the viscosity between oil and water differs, water flows at a greater velocity than oil. At low flow rates, however, this difference may be equalized by the counteraction of gravity which latter seeks to keep the oil/water contact in a horizontal position.

Beyond a "critical production rate" q_c, however, this kind of equilibrium is no longer possible, and a water tongue is formed. Fig. 102 shows this schematically on the flank of an oil reservoir.

When (from the equation $\frac{k_o \mu_w}{k_w \mu_o} = M =$ mobility ratio) M is known on the one hand and when S_{or} is known on the other the problem is, in principle, capable of solution. An integration will

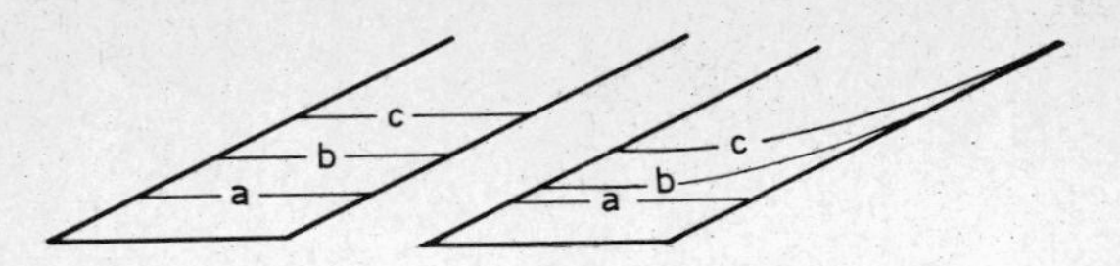

Fig. 102 Horizontal advance of the water/oil contact line from a via b to c (left) and formation of water tongues when the "critical production rate" q_c is exceeded (right). (After *Dietz*[2])

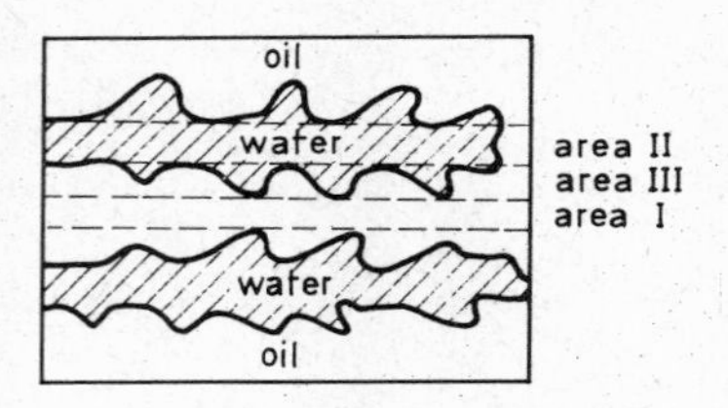

Fig. 103 Let us imagine a cross sectional unit of a reservoir in which sufficient water fingers have formed. We consider them as representative of the entire reservoir. Now if we presume that, originally, the pore space was completely saturated with oil (no connate water present) then Fig. 103 can be divided into three different areas:
1. The area outside the water fingers where oil is the only flowing medium (Area I);
2. The sum of the central areas where water is the only flowing medium without restriction;
3. The area formed by water fingers (so-called protuberances) (Area III) in which oil areas are enclosed. Inside this area oil and water are both immobile. (After *Meurs* et al.[8])

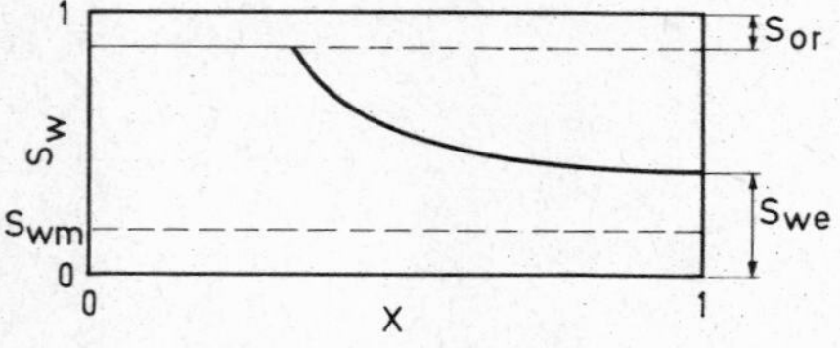

Fig. 104 Schematic representation of water saturation distribution after water breakthrough. (After *Meurs* et al.[8])
X = dimensionless distance along the direction of flow measured from the point of water influx.
l = Length of the reservoir as measured between the point of water influx and the point of liquid outflow.
Q = Cumulative flow or cumulative water influx, respectively, per cross sectional unit area as measured in fractions of pore space

then give the oil recovery as a function of cumulative water inflow (q × t). *Dietz's* theory establishes a relation between effective oil recovery on the one hand, and cumulative water inflow, effective mobility ratio and residual oil saturation on the other.
The displacement theory proposed by *van Meurs* and *van der Poel*[8] is based on the following assumption:
When, in a porous medium, oil is displaced by water (the viscosity of which is much lower) then the oil/water interface will necessarily be unstable and tend to shape numerous finger-like forms (viscous fingering). Tests performed on transparent models confirmed this "viscous fingering" in an impressive manner. *van Meurs* and *van der Poel* based their displacement theory on these observations and developed a formula describing cumulative production as a func-

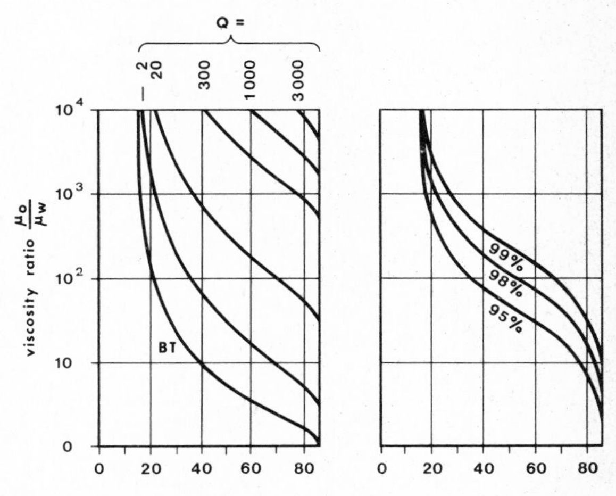

Fig. 105 When the viscosity ratio between oil and water $\frac{\mu_o}{\mu_w} = 100$ then, as Fig. 105 shows, 22 % of the oil in place can be recovered until the water breaks through to the well. Then, if production is continued (for the oil + water produced after water breakthrough) until total wet oil production equals 20 times the pore space available between the edge water line and the well (Q = 20), 70 % of the oil in place are recovered. Under these conditions, however, the water cut would already exceeds 99 % and production would be entirely uneconomic. However, if the viscosity ratio $\frac{\mu_o}{\mu_w} = 1$ (which means that oil and water are of equal viscosity) then all of such oil as is physically recoverable (of the order of 85 % of the oil in place) can be recovered prior to water breakthrough. (After *Meurs* et al.[8])

tion of cumulative water flow on the one hand, and of the oil/water viscosity ratio on the other.
The size of the three areas in Fig. 103 can be described by water and oil saturation values. The immobile water share in Area III is designated as S_{wm} while the immobile oil share in it is designated as S_{or}. Saturation is related to the area unit. Medium water saturation is S_w. The size of Area I equals $1\text{-}S_{or}\text{-}S_w$ while the size of Area II equals $S_w\text{-}S_{wm}$. The schematic representation in Fig. 104 describes saturation distribution in the unit area.
Oil recovery after water breakthrough is designated by the letter R and, like cumulative flow Q, is expressed in fractions of pore space. Fig. 105 shows how strongly the oil recovery varies with the viscosity ratio and with cumulative throughflow Q. Beyond this the right-hand figure represents recovery curves vs. various economical production limits, namely for $f_{we} = 0.99$, 0.98 and 0.95.
Theoretical results were verified by experiments on models. A satisfactory conformity of oil recovery was found to exist in the range of $R = 0.20$ to 0.75. The question still is what importance should be attributed to the quantities S_{or} and S_{wm} for the purposes of production forecasts in water drive reservoirs. As, in the case of highly viscous oils the degree to which S_{or} affects recovery is very small, S_{or} can be presumed to equal 0.15 in the majority of practical cases. And as all model tests made hitherto showed the best conformity with theoretical results to be obtained for $S_{wm} = 0.15$ it is altogether permissible to start out from this value in practice.

3.4 Remarks on the Recovery Factor

From the relative permeability curves it will be obvious (page 82) that oil ceases to flow ($k_{ro} = 0$) when oil saturation drops to a certain value. It is generally believed that this residual oil (S_{or}) is retained in the pores by "capillary forces". There is also a certain water saturation which is kept at the pore walls by capillary forces and thus does not flow. We call this irreducible saturation (S_{wi}) or "connate water" (S_{wc}). In the Hg-injection method it is that Hg-saturation at which no additional Hg can be injected, however great the pressure applied may be. In the relative permeability curve it is that part of the saturation for which $k_{rw} = 0$. Whilst it is easy to explain the irreducible water saturation (since water is the wetting phase and therefore covers the surface of the sand grains like a thin film) it is more difficult to find an explanation for the residual oil saturation. Its amount depends on "pore geometry". One possible way of retaining oil in the pores of the reservoir rocks is schematically shown in Fig. 106. Another probable explanation is the so called "viscous fingering" (Fig. 103, where the residual oil saturation is marked by area III).

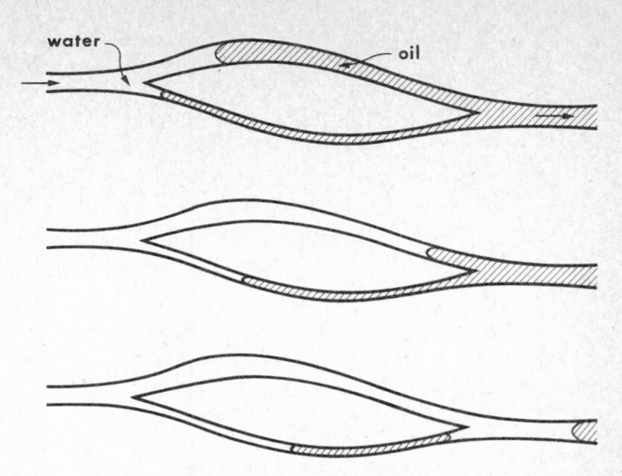

Fig. 106 A large portion of the unrecoverable oil is retained by "capillary forces". In the complicated network of the pore space in reservoir rocks, capillaries with different radii are the rule. These capillary conditions (together with the viscosities of oil and water) are responsible for the residual oil saturation of reservoir rocks. After *Moore* et al.[9])

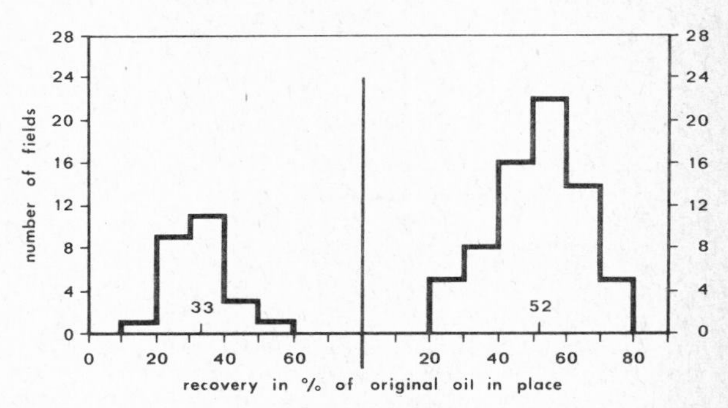

Fig. 107 The recovery factor of a number of US oil fields with a gas drive mechanism is between 10 and 60 %, the average being 33 % (left side). Fields with a water drive have an average a recovery factor of 52 % (right side). (After *Muskat*[10])

This residual oil remains in the reservoir, as it cannot be recovered by normal methods (including water flooding)[14]. In oil-field practice the residual oil saturation S_{or} varies between 20 and 80 %. For the oil-fields in the Federal Republic of Germany it averages some 70 %. The recovery factor (the producible oil as a percentage of the oil in place) is thus about 30 % in this country.

In the case of oil-fields with a natural water drive (or with water flooding) the time required for the production of the recoverable oil is of tremendous importance. The f_w curves (Fig. 94) show the rapid increase of the water cut (f_w) as oil saturation decreases. It soon gains a share of more than 95 % in total flow. This large water cut constitutes a grave technical and economic problem and materially affects the life and, along with this, the ultimate recovery of an oil-field. In frequent cases a water cut of 95 % (or even less) prohibits economic oil recovery.
The f-functions show how strongly f_w (and f_o, respectively) depend on the viscosity ratio between oil and water $\frac{\mu_o}{\mu_w}$.
It will readily be seen from the example of Fig. 108 what influence

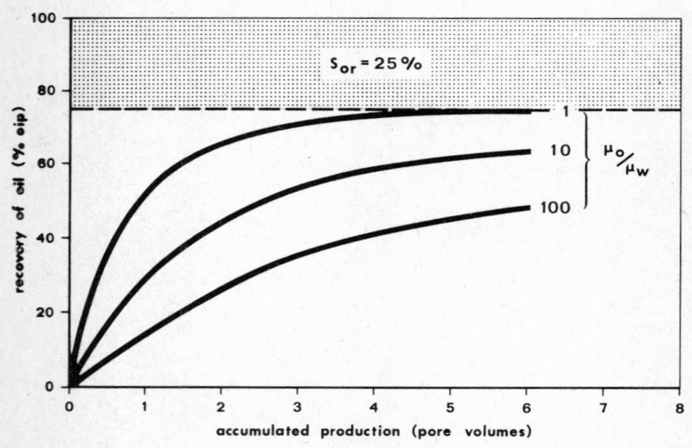

Fig. 108 Influence of the viscosity ratio $\frac{\mu_o}{\mu_w}$ on the recovery of oil. This example shows that 25 % of the oil are retained in the pores "by capillary forces" and cannot be recovered. However, if the viscosity ratio is very low (e.g. $\frac{\mu_o}{\mu_w} = 1$) then, in the above example, all of the recoverable oil is produced when the total flow of oil + water is some 5 times as large as the reservoir's total pore volume or, in other words, when cumulative wet oil production attains a volume of 5 PV. When, on the other hand, the viscosity ratio $\frac{\mu_o}{\mu_w} > 1$, residual oil saturation is not attained until cumulative production exceeds 10 pore volumes. When the viscosity ratio equals 100 this stage is reached much later. (After *Tunn*[14])

this viscosity ratio has on production *time*. As oil viscosity strongly depends on temperature, it is the aim of all thermal procedures to reduce $\frac{\mu_o}{\mu_w}$ by raising the temperature in the reservoir. See the example of Fig. 109.
Such high water cuts are a paramount economic factor which, in the end, determines the lifetime of the field in question (refer to

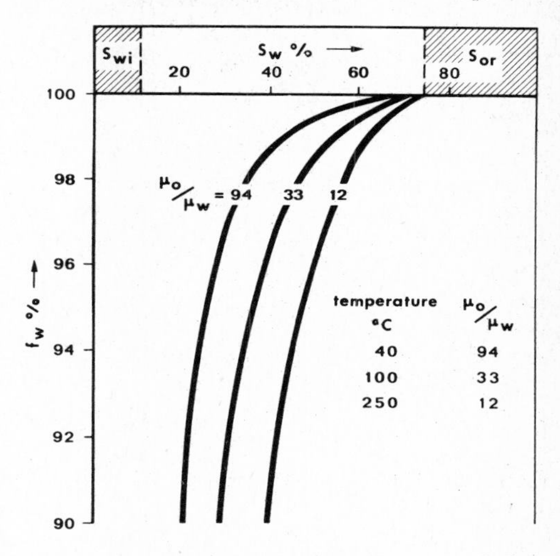

Fig. 109 represents the share of the water f_w in total flow (the water cut in % for values between 90 and 100 %) at viscosity ratios $\frac{\mu_o}{\mu_w}$ of 94, of 33 and of 12. (When the oil is very viscous and $\frac{\mu_o}{\mu_w}$ equals, for instance, 94, the water cut amounts to 90 % when water saturation S_w is merely 20 %. When the oil is less viscous and $\frac{\mu_o}{\mu_w}$ equals a mere 12 a water cut of 90 % is not reached until water saturation amounts to 40 %.) In the figure, the relation between temperature and $\frac{\mu_o}{\mu_w}$ is illustrated by 3 numerical examples. The viscosity ratio drops from 94 (at 40 °C) to 12 (at 250 °C). While, at a viscosity ratio $\frac{\mu_o}{\mu_w} = 94$ and at a water and oil saturation of 50 %, flow volume is composed of 99.5 % water and just 0.5 % oil. This high water cut would be reduced to 96 % if, under the same conditions of saturation, the viscosity ratio could be reduced to 12 by a substantial rise of temperature

page 101). Water cut will inevitably reach values of over 95 % in the last years of the field. Whether or not such a high water cut still permits economic production with all the gathering and dehydration problems involved, depends entirely on local conditions.

For a long time many authors have tried to determine, at least from a theoretical point of view, what should be done to mobilize the residual oil as well. In this connection *Taber*[13] proved that this can be done when a certain critical value of the formula $\Delta P/L \cdot \sigma$ is exceeded. ($\Delta P/L$ being a pressure gradient P over a length L, and σ being the interfacial tension between oil and water).

Although this critical value is a specific property of each reservoir rock, it seems to be of a certain order for most rocks. It is usually about five, when the pressure gradient is expressed in psi, the length in ft and the interfacial tension in dyn/cm.

When this critical value is exceeded the residual oil content is a function of $\Delta P/L \cdot \sigma$. At high values for this term (for instance 100 times the critical value), all of the residual oil is practically recovered.

The relation between this quotient and the quantity of residual oil recovered is shown in Fig. 110. As the pressure gradient in the reservoir will seldom exceed 0.1 bar/m (0.5 psi/ft) this means in practice that the critical value will not be reached until interfacial

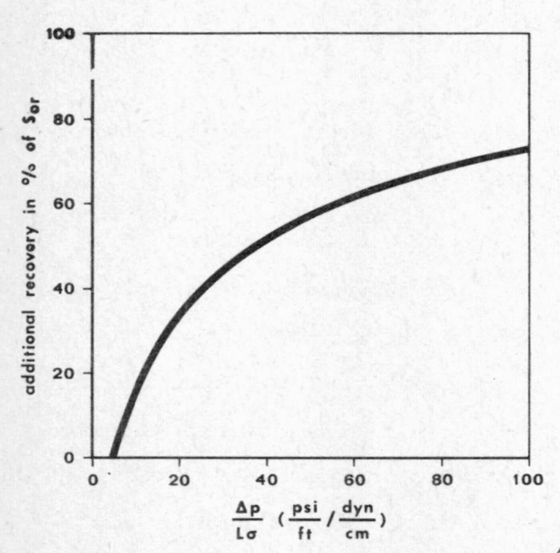

Fig. 110 When a certain critical value of the formula $\frac{\Delta p}{L\sigma}$ is exceeded the residual oil saturation can be mobilized and large portions of this (otherwise not recoverable) oil can be produced. (After *Taber*[13])

tensions of the order of 10^{-2} dyn/cm are attained and that all of the residual oil will be recovered when interfacial tension is of the order of 10^{-4} dyn/cm.

3.5 Material Balance

The complicated play of the forces acting in the pores of the reservoir rocks as soon as the existing equilibrium is disturbed (by the start of production) may be demonstrated in its full complexity by material balance calculations. They are based on the principle that the production of oil and gas as a physical process obeys the law of mass conservation. Therefore an equilibrium exists at all times between the fluids produced on the one hand, and the material (including pore compressibility) replacing this production in the reservoir on the other. For the pore space left behind after the withdrawal of oil and gas for production purposes naturally does not remain empty. The withdrawal leads at first to a decline in reservoir pressure. This results in an expansion of the hydrocarbons which have been left in the reservoir rock and often water enters the reservoir originally filled by oil and gas.
The equilibrium between the produced and the replacing material is expressed by the "material balance". Since we have seen that the physical condition of water, oil and gas in the reservoir differs considerably from their state on the surface, all volumes in the material balance should be converted to uniform physical definitions. The normal reference values are the pressure and temperature prevailing in the reservoir at the time when the material balance equation is formulated. The conversion factors are obtained from the PVT analysis.
But also in field work, material balance calculations have become an indispensable tool for the following reasons:

1. by using them the initial oil and gas volume may be calculated (oil and gas in place).
2. In fields with active water drive, the volume of inflowing edge water can be determined. This enables the strength of the edge water drive to be measured.
3. They enable predictions to be made on the reservoir performance at different production rates.

The principle of such calculations will be explained by a simple schematic example, namely an undersaturated oil reservoir without an edge water drive (see Fig. 111 and Appendix L). This example will also serve to explain how the oil in place is determined.
If the reservoir pressure falls below the bubble point pressure (or in fields with gas caps), the volume of the free gas should be taken into account. The principle is completely identical, however, with

the above example; the only unknown parameter is N which may then be calculated.
For gasfields without a water drive the material balance equation may be formulated as follows:

$$G_p = G\,(1 - \frac{p_n\,z_i}{p_i\,z_n})$$

G – gas initially in place
G_p – cumulative production
p_i – initial reservoir pressure
p_n – reservoir pressure at G_p
z_i – initial real gas factor (at p_i)
z_n – real gas factor (at p_n)

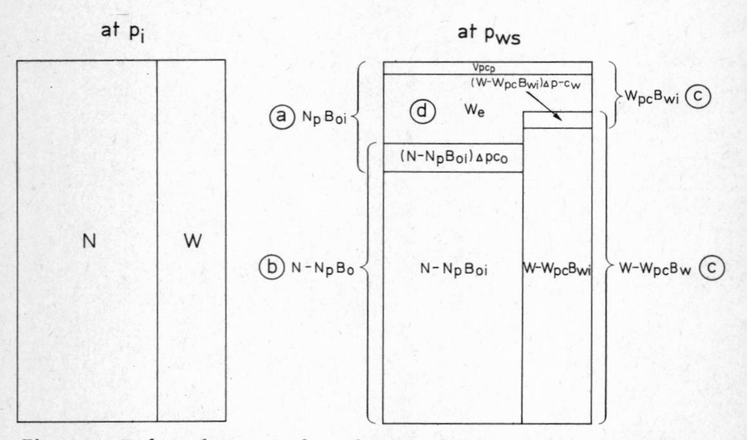

Fig. 111 Before the start of production the pore volume is filled with oil (N) and connate water (W) at an initial pressure p_i. Expressed by an equation: $V_p = N + W$ (m^3).
After N_p m^3 of oil and W_{pc} m^3 of connate water have been produced the pressure is assumed to have fallen by Δp to p_{ws} (bar). The content of the pore volume has thereby changed as follows:
a) the original oil volume N was reduced by the oil production N_p, which means, under reservoir conditions, $N_p\,B_o$.
b) the remaining oil volume $(N - N_p\,B_o)$, however, as a result of the pressure decline Δp has expanded by $(N - N_p\,B_o)\,\Delta p c_o$.
c) the same applies to connate water.
d) when taking also the reduction of the pore volume by the pore compressibility $V_p\,c_p$ then the volume "evacuated" by production has been refilled by water influx W_e.
This is explained in some more detail on the example of a German oilfield in Appendix P

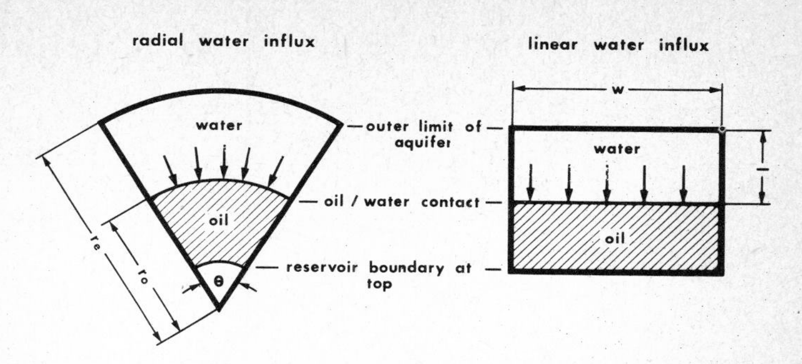

Fig. 112 Schematic representation of radial and linear water influx into an oil reservoir

In fields with water drive the equation contains however two unknown values: the initial oil volume N and the water influx into the reservoir W_e. If one of these factors is known the other may be calculated. (It is possible to determine the initial oil volume N with the aid of volumetric reserve calculations or to estimate the water influx by observations in wells close to the edge water line). Or several material balance calculations are made at greater intervals, which will then result in approximations for N and W_e (least squares method).

The extended material balance equation
Since the work of *R. J. Schilthuis* (1936) and the studies by *A. F. van Everdingen* and *W. Hurst*[3] the material balance equation may be applied to manifold problems of many different water drive reservoirs, even if the original oil in place is not known.
When applied to water drive reservoirs, the material balance equation is extended by the water encroachment factor W_e, the cumulative water influx, and is now written in the general form

$$Q = N \times a + W_e \tag{1}$$

The equation says that the cumulative production (Q) equals the expansion of the original oil in place + the cumulative water influx (a is an expansion factor).
In most water drive reservoirs, the water influx W_e is caused by the expansion of water in the aquifer (= waterbearing formation) due to pressure decline Δp in the reservoir. The amount of inflowing water depends on the water volume, the geometry of the aquifer and the physical rock and water parameters. Only in a few cases can a linear relationship be established between W_e and Δp; this is possible only in highly permeable formations, where any pressure reduction is accompanied by an almost simultaneous pressure

equalization over the whole formation. By solving the partial differential equation of hydrodynamics, the so-called "equation of continuity", *van Everdingen* & *Hurst* have derived a relationship which makes it possible to determine the water influx from a number of parameters. In generalized form, this equation reads

$$W_e = U \times W_D. \qquad (2)$$

It gives the cumulative water influx per unit of pressure decline (= 1 bar).
U is an aquifer constant and W_D a dimensionless water influx quantity, which represents the water influx related to the unit of pressure decline as a function of the dimensionless time parameter t_D. U and W_D furthermore depend on the geometry of the aquifer, which determines whether the influx is *radial or linear* (see Fig. 112), and on the size of the aquifer, which may be *limited or infinite.*
Actual values for the unit function W_D can either be taken from tables or graphs of the relevant literature.
With the aid of the aquifer theory developed by *van Everdingen* & *Hurst,* it is now possible to use the material balance equation [equation (1)] in solving various problems, e.g.:

1. Determination of oil in place N
Equation (1) can be solved with respect to the two unknown quantities N and W_e by "trial and error", if the production history is known.
2. *Forecast of future production performance*
If N is known, the reservoir parameters can be determined from the production history. These parameters make it possible to forecast future production performance (production rate and pressure).
3. *Estimation of recovery*
Since the anticipated cumulative water influx can be calculated when assuming a certain future production performance, it is possible to estimate the primary recovery according to the displacement theory. An aquifer analysis will then show if the aquifer is large enough to permit maximum recovery. This point is very important, as it governs the decision whether to introduce secondary recovery measures or not.

Supplement: The average reservoir pressure
In material balance calculations, an average reservoir pressure for the whole field has to be assumed. In practice, this encounters certain difficulties, since the pressures measured in individual wells may differ a lot from each other. In the vicinity of wells with high production rates, reservoir pressures will always be lower than in wells which were closed in for prolonged periods; mostly pressures

near the edge water are higher than those far from it, etc.; pressure equalization over the whole field would normally last very long and could be definitely achieved only by closing down the field.
It is therefore necessary to use average values, which can be cal-

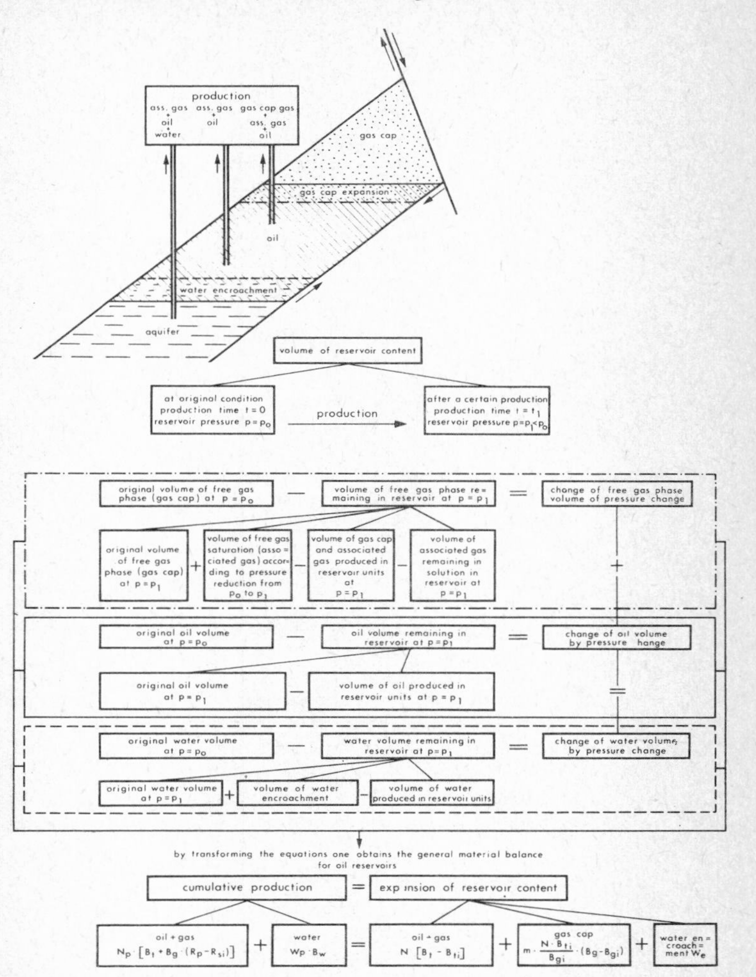

Fig. 113 Schematic Representation of "Material Balance" of an Oil Reservoir. (From *Jandl, A.:* Reservoir Engineering. "Unser Betrieb", Deilmann 1975)

culated from the pressures measured in individual wells in three different ways:

1. Formation of the arithmetic mean from the sum of all wells.
2. The "drainage area" of a well is determined in m^2 as accurately as possible, this area is multiplied by the closed-in pressure recorded for each well and the arithmetic mean is then formed from the sum of the products.
3. One takes into account also the thickness of the reservoir rock and determines the "drainage volume" of each well in m^3 by using isopach maps. This is naturally the most exact method. It is, however, time-consuming and it divides the field into the "Associated Reservoir Volumes (ARV)" pertaining to the individual wells. These volumes (which do not necessarily correspond to the drainage areas) are considered as separate reservoir units, in which the average reservoir pressure as an unknown quantity is related to the oil in place[5].

3.6 Reservoir Models and Simulations

3.6.1 Reservoir Models

The flow of fluids through permeable rocks has been studied on rock samples since the early days of the oil industry. Water supply problems of the French city of Dijon led to flow experiments in 1859 by M. Darcy which formed the basis of permeability considerations. Later, similar experiments were carried out by the oil producing industry. These experiments used water flooding to study the encroachment of water, but were then extended to all flow processes. The rock specimens became more and more extensive and finally led to real scale models of oil and gas reservoirs. The flow of gas, oil and water could be changed and adapted to the natural environment. It was a demonstration for empirical observation, since the properties of the reservoir rock and the transparency of the whole model led to a near perfect artificial simulation. The observation of the phase behaviour in a high pressure autoclave may also be regarded as a substitute for a reservoir.

Reservoir problems defying an empirical solution can be tackled with the help of an electrical model. The fundamental concept underlying the latter is that of the analogy of the laws formulated by *Darcy* and *Ohm*.

As current intensity I equals voltage drop ΔE divided by resistance R or, in a formula $I = \frac{\Delta E}{R}$, and as resistance R is proportional to the length of the electrical conductor and inversely proportional to its cross section or, in a formula $R = \frac{\varrho \times L}{A}$, ($\varrho$ being the resistivity of the conductor in question), these relations may be summarized as follows:

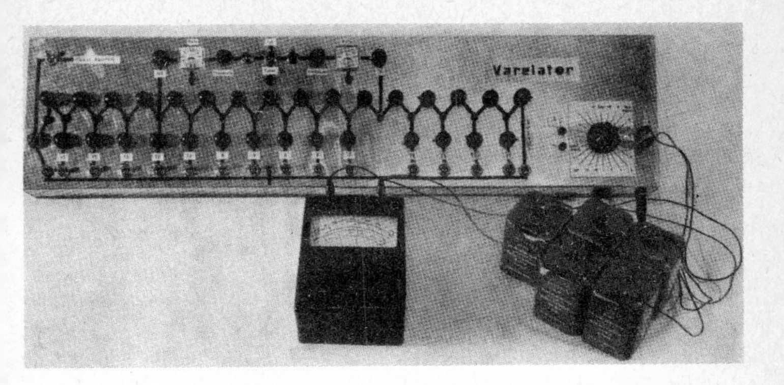

Fig. 114 shows an electrical model of a small oil-field in Northern Germany. It is a monocline structure which was drilled by only one row of wells at the top of the structure. There is no water drive and two wells were planned as water injection wells

$$I = \frac{\Delta E}{R} = \frac{A \times \Delta E}{\varrho \times L} = \frac{\Delta E}{p \times \frac{L}{A}}$$

The analogy with Darcy's Law is fairly obvious:

$$q = \frac{k \times A \times \Delta p}{\mu \times L} = \frac{\Delta p}{\frac{\mu}{k} \times \frac{L}{A}}$$

Current intensity I	corresponds to q	(flow rate)
Voltage drop ΔE	corresponds to Δp	(pressure difference)
Conductivity $\frac{1}{\varrho}$	corresponds to $\frac{k}{\mu}$	(mobility)
$\frac{1}{R} = \frac{A}{\varrho \times L}$	corresponds to $\frac{k \times A}{\mu \times L} = \frac{q}{\Delta p}$	(productivity index)

The feature of major importance to production problems in oil reservoirs is the analogy between conductance $\frac{1}{R}$ on the one hand and productivity $\frac{q}{\Delta p}$ on the other. This analogy also applies to radial flow. It permits the formula for radial flow on page 79 to be written as follows in electrical analogy:

$$I = \frac{2\pi h \Delta E}{p \ln (r_e/r_w)}$$

This formula would, for instance, be applicable in the following simplest reservoir model: a tank (without a roof) consisting of a metal wall and non-conductive bottom (glass or plastic) is filled with a conductive liquid (brine). The wall serves as a positive electrode. For the negative electrode a metal rod is inserted vertically in the liquid in the centre of the tank. The current now flows radially from the wall (border of the reservoir) to the central

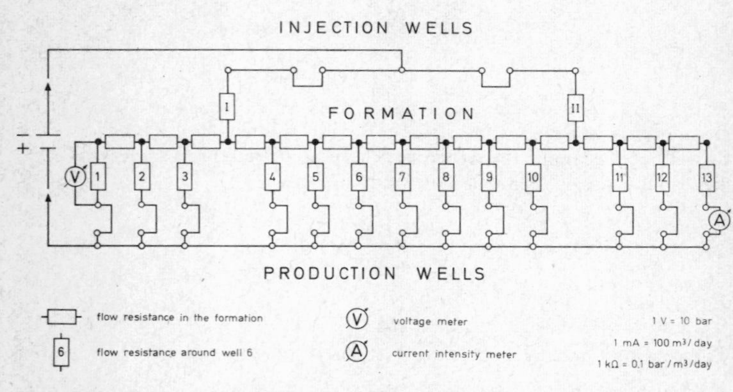

Fig. 115 shows a simplified schematic representation of the electrical model. By using it the production and pressure performance of each well can be easily determined. In the model the wells are represented by electrical resistances corresponding to $1/q/\Delta p$ in the immediate vicinity of the wells. This is the productivity index PI which is measured for each well. The electrical resistance is then adjusted accordingly. The electrical resistances applied between the wells correspond to the flow resistances in the formation. The example shows how in well 1 the voltage (= pressure) and in well 13 the current intensity (= rate of production) are measured

---

Fig. 116 The choice of the grid layout for mathematical reservoir simulators is determined by the type of problem to be studied and the geometrical detail required. If the behaviour of bottom water is to be studied the grid may consist of vertical cylinders with different radii concentrated around the well and horizontally cut into "slices" (example b.). In other cases the field is divided into regular blocks in horizontal and vertical direction as in the study for thermal recovery in a German field (c). A Bavarian field trapped by an antithetic fault line is shown under a

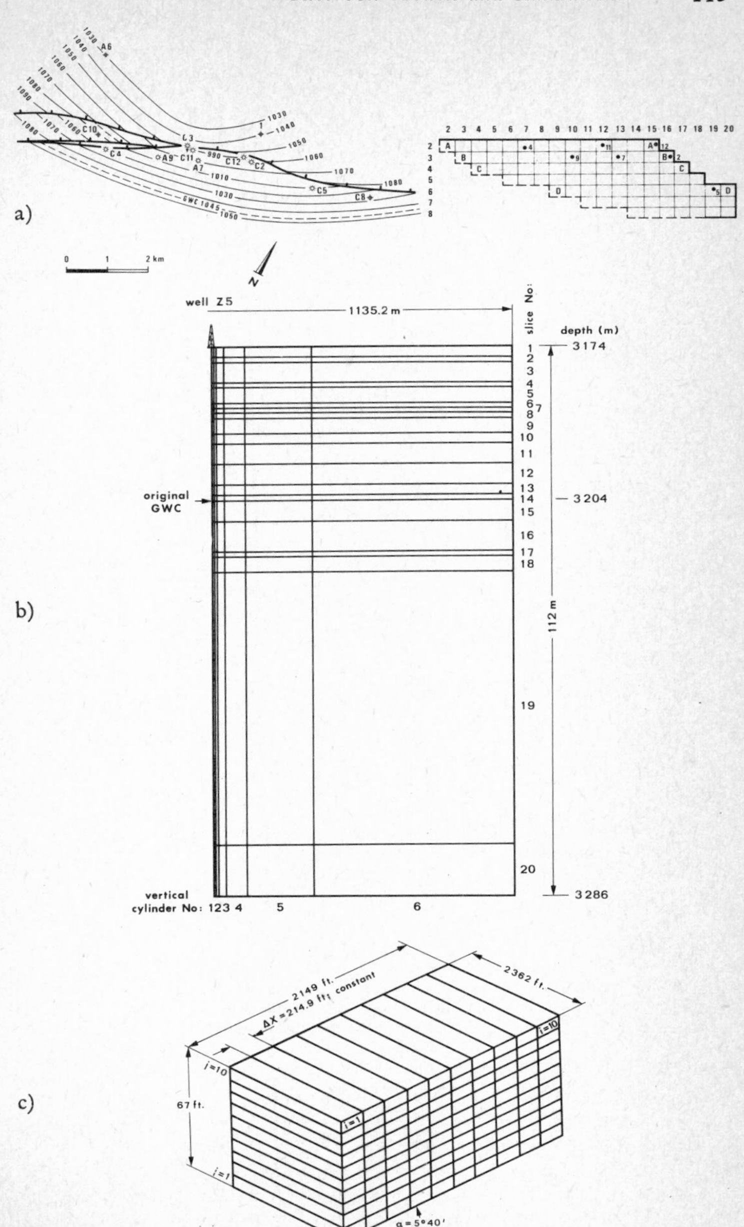
a)
b)
c)
0 1 2 km
N
well Z5
1135.2 m
slice No:
depth (m)
3174
3204
3286
112 m
original GWC
vertical cylinder No: 1 2 3 4 5 6
2149 ft.
ΔX = 214.9 ft; constant
2362 ft.
67 ft.
α = 5°40'

electrode (well). The voltage drop (pressure difference) and the current intensity (production rate) may now be measured (see Figures 114 and 115).

3.6.2 Mathematical Reservoir Simulation

A mathematical reservoir simulator is a computer program which calculates the behaviour of an oil- or gas-field by representing the above mentioned flow phenomena by differential equations and solving them numerically.
For this purpose the reservoir is divided into blocks. The size of these blocks depends on the quantity of available data (and wells), the type of the problem and the required accuracy of the overall simulation. Each block can be represented by a finite-difference equation, which takes into account the flow of the different media (or the phases). A two-dimensional model, for instance, can be used where vertical equilibrium exists. The time is accordingly divided into a number of small time steps. For each step the flow equations must be solved simultaneously.
Since one of the most frequent targets of the computer program is to forecast the reservoir performance under different plans of operation, an attempt is made to match the available production data by changing reservoir parameters and/or fluid parameters which are believed to influence the reservoir behaviour in the expected manner.

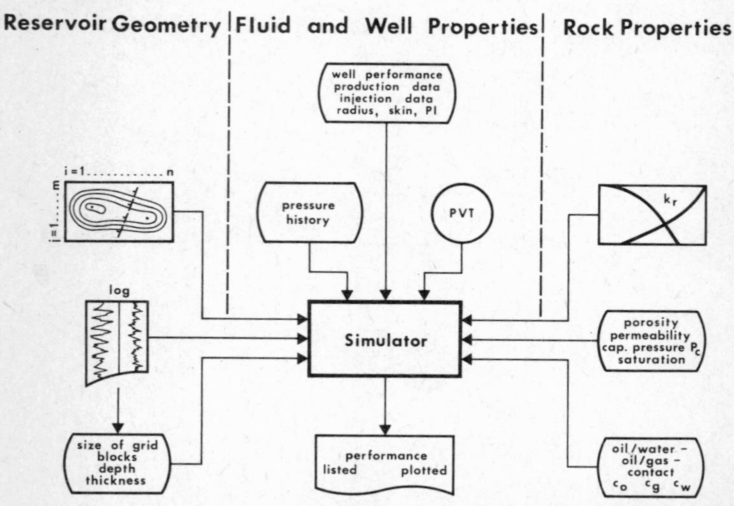

Fig. 117 Schematic representation of data flow in a mathematical reservoir simulation model. (Somewhat changed after *Schröckenfuchs* et al.[12])

At the start of a time step the pressure and the saturation are known in the form of arrays which contain one average value per block. Since pressure potentials govern the fractional flow of the different phases this leads to a changed situation at the end of a time step. In principle the procedure is a material balance calculation as represented in Fig. 111 which is then repeated in numerous steps for each block. There are a number of input data for the simulation program which are schematically represented in Fig. 117. The model itself is determined by the type of the problem: there are trend studies (that is a simplified representation showing the effect of parameter changes) and there are models for obtaining accurate results through as much detail as possible. A great number of models are available and are generally supplied by service companies.

"An appropriate reservoir simulation model can be used to understand problems associated with a given process and to indicate remedies for these problems. The simulation model can also be used to determine the sensitivity of performance to certain parameters, thus providing information necessary to optimize a process" (Ninth World Petroleum Congress, Tokyo 1975).

Bibliography

[1] *Buckley, S. E., M. C. Leverett:* Mechanism of Fluid Displacement in Sands, Trans AIME, 1941.

[2] *Dietz, D. N.:* A Theoretical Approach to the Problem of Encroaching and By-Passing Edge Water. Proc. kon. Nederl. Akad. Wet., Ser. B, 1., Amsterdam 1953.

[3] *Everdingen, A. F. van, W. Hurst:* The Application of the Laplace Transformations to Flow Problems in Reservoirs. J. Petrol Technl. London, AIME, Petrol Trans., 168, New York 1949.

[4] *Hooykaas, J.:* Appraisal of Marsal's Extrapolation Method of Establishing Oil Recovery. AIME, SPE 919, 1964.

[5] *Horner, D. R.:* Average Reservoir Pressure. Proc. Fourth World Petrol. Congr., Sect. II/E, Rom 1955.

[6] *Leverett, M. C., W. B. Lewis:* Steady Flow of Gas-Oil-Water Mixtures through Unconsolidated Sands, Trans. AIME, 1941.

[7] *Marsal, D.:* Die Berechnung der Verwässerung von Erdöllagerstätten in eindimensionaler Näherung. Erdöl und Kohle, 1957.

[8] *Meurs, P. van, C. van der Poel:* A Theoretical Description of Water-Drive Process Involving Viscous Fingering. Trans AIME, 213, S. 103, New York 1958.

[9] *Moore, T. F., R. L. Slobold:* Producer's Monthly, 1956.

[10] *Muskat, M.:* The Flow of homogeneous Fluids through porous Media. New York–London 1937.

[11] *Pirson, S. J.:* Oil Res. Eng., New York–London 1958.

[12] *Schröckenfuchs, G., St. Logigan, J. Spitzl:* Erdöl-Erdgas-Zschr., Sept. 1973.

[13] *Taber, J. J.:* Dynamic and Static Forces Required to Remove a Discontinuous Oil Phase from Porous Media Containing Both Oil and Water. Soc. of Petr. Eng. J. (March 1969) 3–12.

[14] *Tunn, W.:* Gedanken über Einsatzmöglichkeiten von Tensiden zur Erhöhung der Ausbeute von Erdöllagerstätten. Erdöl-Erdgas-Zschr., Heft 2, 1974.

[15] *Welge, H. J.:* Simplified Method of Computing Oil Recovery by Gas and Water Drive. Trans AIME, 1952.

4 The well — our point of observation

Those parts of an oil or gas reservoir available for direct observation are restricted to rather insignificant random spots: the cores taken from the reservoir rock and the cuttings contained in the mud. A welcome supplement to these observations and measurements on samples is offered by geophysical downhole logging, part of which has a significant "penetration depth" beyond the wall of the bore hole. A still "deeper" insight into the reservoir as a whole is provided by theoretical and empirical investigations on the flow of liquids in permeable rock as a function of pressure, saturation, permeability and fluid properties.

4.1 Pressure measurements

Reservoir energy finds its most obvious expression in reservoir pressure. To the reservoir engineer, this pressure is the basic parameter and the most important variable for evaluating the present and future reservoir performance.
As is generally known, the static equilibrium prevailing in the reservoir prior to the start of production is disturbed by lowering the mud pressure below the reservoir pressure. This "point of disturbance", the well, is at the same time our only point of observation. Here we carry out those measurements from which we draw conclusions with respect to what is going on in the reservoir.

4.1.1 Methods and devices

It is difficult (and frequently even impossible) to draw conclusions from the pressure measured at the wellhead about the pressure prevailing at the sand face. Therefore it is accepted practice to measure the reservoir pressure by instruments which, as a rule, are lowered down the hole.
The most common pressure gauges make use of the following principles:
As in a barometer, pressure metering is based on the expansion or contraction of a metallic element, usually a spiral bourdon tube. This tube is provided with a recording pin which transmits the upward and downward movements of the tube to a recording drum kept in slow and steady rotation by a clock mechanism. In other instruments the metallic sensitive element consists of bellows made of a beryllium-bronze alloy. Still other instruments are equipped with simple rectangular metal receptacles at the top of which a capillary glass tube is attached.
Instruments of a somewhat different type are based on the expansion or contraction of a metal spring attached to a movable

piston. The piston moves with changing pressure and records it on a drum turned by clockwork. The result is a time/pressure diagram which allows pressure readings at any depth desired. The instrument in widest use in the Federal Republic of Germany is the so-called *Hügel* instrument, which reduces the frictional effects of the piston by a strong clockwork rotation. By a computer correction programme, Hügel achieves as much precision as other highly sensitive instruments such as the Sperry Sun or Maihak gauges.

The diameter of these instruments is sufficiently small to be run into any well using wire and stuffing box. The measuring vehicle (usually of special design) is equipped with drum and depth counter. The instruments are calibrated by means of a piston manometer both at room temperature and reservoir temperature. For these instruments twelve day clocks are available, so the pressure performance of a well can be measured over quite an extensive period.

The Maihak Company of Hamburg has developed a pressure gauge consisting of a so-called "transmitter", which is attached to the tubing string and remains in the well, so that measurements can be made at any time. In the "transmitter" exposed to the reservoir pressure, a fine string is caused to oscillate by an electric pulse. The vibration frequency of this string depends on the pressure prevailing and is transmitted to the surface by a cable. A reference string installed aboveground which vibrates at the same frequency as the subsurface transmitter permits an easy reading of the pressure.

A different type of pressure gauge permitting continuous reading is Ball's "pressure sentry". It consists of a Bourdon tube, from which the pressure signal is transmitted electrically to the surface.

The Sperry-Sun pressure recorder is a recent development based on proved and accepted measuring principles. The pressure-sensitive element is a Bourdon-tube whose position is electronically checked at regular intervals. Recording is made either on a metal sheet (turned by clockwork) or via a cable at the surface. This method is very accurate since it is not affected by frictional forces.

An interesting new development is the Hewlett-Packard pressure recorder. In this case the pressure-sensitive element is an oscillating crystal whose resonance frequency changes with pressure. Recording is made at the surface via a cable. This instrument is probably the most sensitive pressure recorder currently available.

In pumping wells, another instrument, the "sonic depth finder", has been successfully applied. It determines the liquid level in the annulars between tubing and casing. The pressure is then computed from the specific gravity of the oil or water. Pressure measurements in pumping wells are of particular importance to the production engineer, as they give him a clue to the production methods to be applied (dimensioning of the pumps and the flow string, length of stroke, number of strokes etc. suiting the optimum delivery rate of the reservoir).

In flowing wells there are three kinds of measurements usually made:

Interval measurements: the pressure in the tubing is measured at different depths from the surface down to the bottom. This pressure/depth diagram allows valuable conclusions to be drawn with respect to the contents of the hole. From the "pressure gradient" (usually the pressure difference per 10 m) the specific weight is derived; as a rule, this offers a fairly reliable picture of conditions within the well (for instance of the depth in the tubing, where gas is liberated).

Flowing pressure: during production the pressure bomb remains in the well and the gauge measures the pressure (p_f) at different rates of production. The difference between flowing pressure and shut-in pressure (frequently called "draw down") is a measure of rock permeability.

Shut-in pressure: the pressure bomb remains in the hole when production is suspended. The shut-in pressure (p_s) equals reservoir pressure when the pressure is stabilized. When – as is usually the case – the flowing pressure is measured and the well is then shut in, afterwards the pressure will build up. The shape of these pressure build-up curves permits important conclusions to be drawn.

4.1.2 Capacity of a well

When an oil or gas well is completed one of the first problems confronting the petroleum engineer is to determine its capacity (productivity, potential), i.e. the volume of oil or gas that can be produced per unit of time (hourly or daily). The well's delivery rate, however, depends on a number of factors part of which are mechanical (diameter of the bore hole, cross section of tubings, technical equipment of pumping wells etc.) while others are physical (viscosity, pressure, rock properties etc.)

4.1.2.1 The productivity index (PI)

The introduction of pressure measurements makes it possible to determine the capacity of a well without actual open flow which might damage the reservoir. This may be done using the "productivity index" (PI).

The productivity index is the production rate per pressure unit. The production rate depends on the pressure difference between shut-in pressure and flowing pressure: (p_{ws}–p_{wf} = "drawn down"). Whilst the pressures are measured in the wellbore opposite the pay horizon, the quantities (oil + water) are measured under surface conditions. If, for instance, in a given field reservoir pressure (p_{ws}) is 100 bar and if the well produces 45 m^3 of tank oil a day and if,

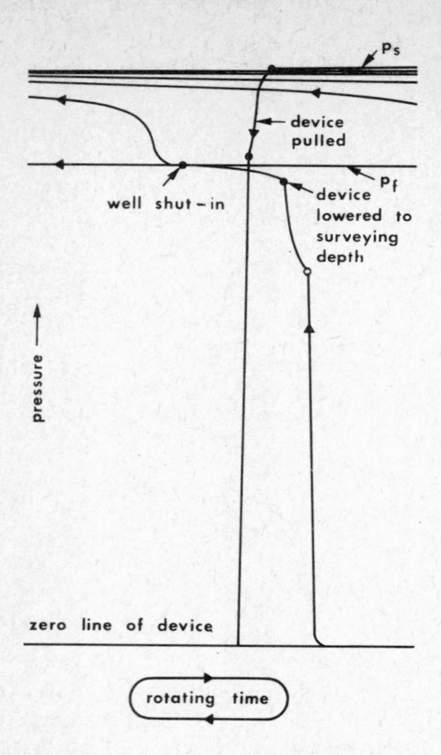

Fig. 118 Measuring Diagram of a Hügel Pressure Bomb. It shows the pressure rise while the instrument is lowered into the well. When the desired depth is reached, flowing pressure p_f is recorded (horizontal line). Afterwards the well is shut-in and the pressure builds up to p_s. (The rotation of the writing pen causes the lines, interrupted at the left, to be continued at the right.)

at this rate of delivery, flowing pressure p_{wf} is 90 bar (draw-down 10 bar), then the productivity index PI amounts to (45 : 10 =) 4.5. Or, in other words, the well produces 4.5 m³ a day per bar of pressure difference. In oil-field practice the productivity index is used frequently since it gives a first rough feature of a well's behaviour.

From this productivity index the theoretical or "absolute" capacity of a well can be derived by multiplying PI and p_{ws}. In the above example, the capacity would be (100 × 4.5 =) 450 m³ a day.

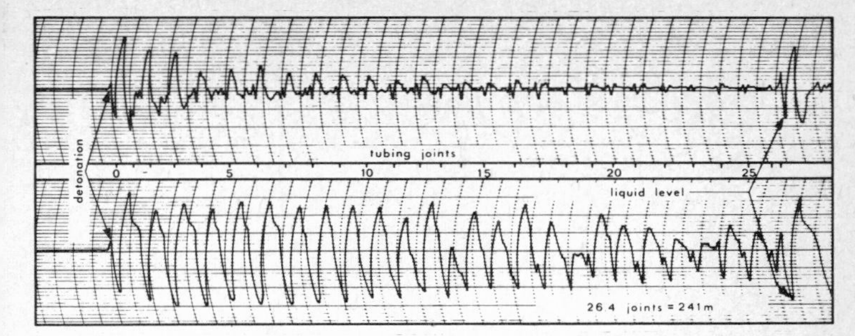

Fig. 119 In pumping wells the "sonic depth finder" is applied successfully. With the help of a small cartridge, a detonation wave is generated at the well head and reflected by the tubing joints and finally by the liquid level. By means of a "seismograph" the waves are recorded on the surface. As the individual depths of the joints and their number are known, the height of the liquid level can be easily determined. The figure shows on the right side the complete reflexion at the liquid level an minor reflexions at each tubing joint. On the left side the reflexions are enhanced by a more sensitive adjustment

To determine the productivity index the well is produced at various rates and the flowing pressures are recorded. The productivity index then results from the formula:

$$PI = \frac{q}{p_{ws} - p_{wf}}$$

(in which q is measured under surface conditions). See Fig. 120.
Since the productivity index may be derived from the radial flow formula (page 79) too:

$$PI = \frac{q}{p_{ws} - p_{wf}} = \frac{2 \times h \times B_0}{\mu \times \ln r_e/r_w}$$

it can be concluded theoretically that PI is not constant, since
a) r_e is a function of time and therefore not constant
b) the formula applies to steady state flow only
c) at high rates of production, flow within the pores becomes turbulent
d) in the case of gas liberation, relative oil permeability and viscosity are reduced
e) relative oil permeability is reduced, too, in case of non-zero water cut.
For a comparison of wells with different pay thickness the "specific productivity index PI_s" is an excellent device. This is obtained by

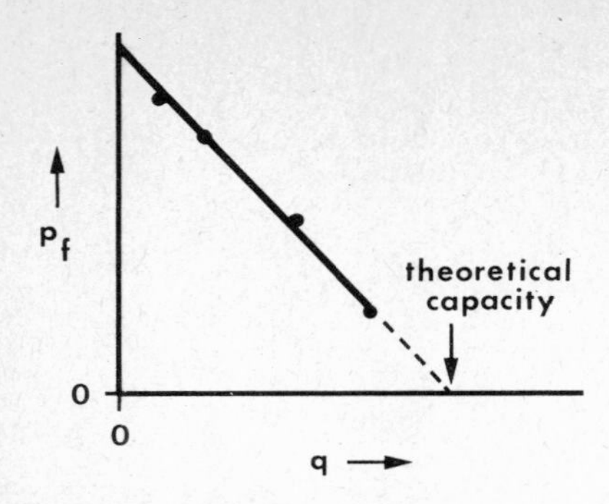

Fig. 120 The introduction of pressure measurements allows one to calculate the capacity of a well without producing it open hole. When plotting the flowing pressure (p_{wf}) against the production rate (q) the capacity of the well may be determined graphically. If a straight line results, the productivity index (PI) remains "constant"

dividing PI by the net thickness of the reservoir rock. If, in the above example, net thickness equals 10 m then $PI_s = 0.45\ m^2/d \cdot bar \cdot m$.
Conversely it is possible to define the "injectivity index" for injection wells. It is the volume of liquid (in m^3) that can be injected daily into the reservoir at an injection pressure exceeding the reservoir pressure by 1 bar.

4.1.2.2 Testing the capacity of gas wells

An exact determination of the capacity of gas wells is of considerable practical importance for the petroleum engineer as it governs the fulfilment of long-term supply contracts. Both the venerable rules of thumb (such as the one that daily production equals 1/10000 of the gas in place) and the original "open flow tests" do not satisfy modern engineering requirements any more. (The so-called "open flow test" measured the volume produced at fully opened wells. Apart from the enormous waste of gas a damage to the reservoir was frequently inevitable.)
The relation between the production rate of a well and the respective flowing pressure may be expressed empirically by the formula:

$$q = C\ (p_{ws}^2 - p_{wf}^2)^n$$

in which

p_{ws} = pressure in the reservoir (stabilized shut-in pressure)
p_{wf} = flowing pressure in the well
q = production rate

The constants C and n are functions of reservoir parameters like permeability, temperature, viscosity etc. Since the relationships are not sufficiently known the capacity equation constants are established empirically by production tests.
If the gas flow were to follow Darcy's Law exactly, then n would equal 1. It has been found, however, that, in actual gas production this law never does apply exactly due to turbulence in the pores. It follows that n usually is smaller than 1. Many methods have been developed to adapt the flow equation to this non-Darcy flow behaviour of the gas. After production tests n is found empirically in the inclination of the straight line resulting when log $(p_{sw}^2 - p_{wf}^2)$ is plotted against log q (see Fig. 121).

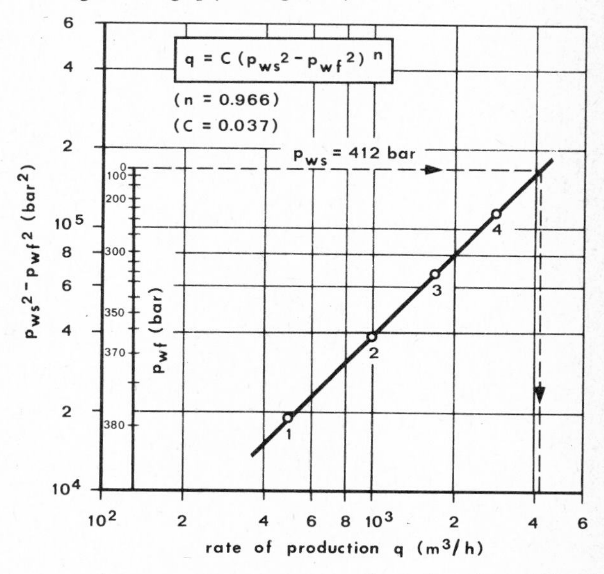

Fig. 121 "Back Pressure Test" in a gas well. At first the stabilized shut-in pressure (p_{ws}) has to be determined. Then two to four stabilized flow rates (q) (using beans of different sizes) with the relevant flowing pressure (p_{wf}) have to be measured. The variable ($p_{ws}^2 - p_{wf}^2$) is plotted on the ordinate vs. the production rate (q) on the abszissa logarithmically. (In order to make the diagram somewhat handier to use the flowing pressure (p_{wf}) is often plotted additionally.) In the example above the "draw-down" ($p_{ws} - p_{wf}$) is 412 bar at a flowing pressure (p_{wf}) of zero bar. The absolute or "theoretical capacity" of this well is 4100 m³/h. The inclination of the straight line leads to tg $\alpha = n$; in the above case $n = 0.966$. C is then derived from the capacity equation (.037 in the above example)

In order to obtain stabilized flow rates and flowing pressures extended periods of time and large gas volumes are needed especially when the permeability of the reservoir rock is low.
The capacity constant C is not a real constant but changes with the duration of the production period. Flow conditions around the well are in the beginning non-steady and become gradually steady or, at least, semi-steady. The isochronal test method as developed by *Cullender* (1955) makes use of this fact, See Fig. 122 and

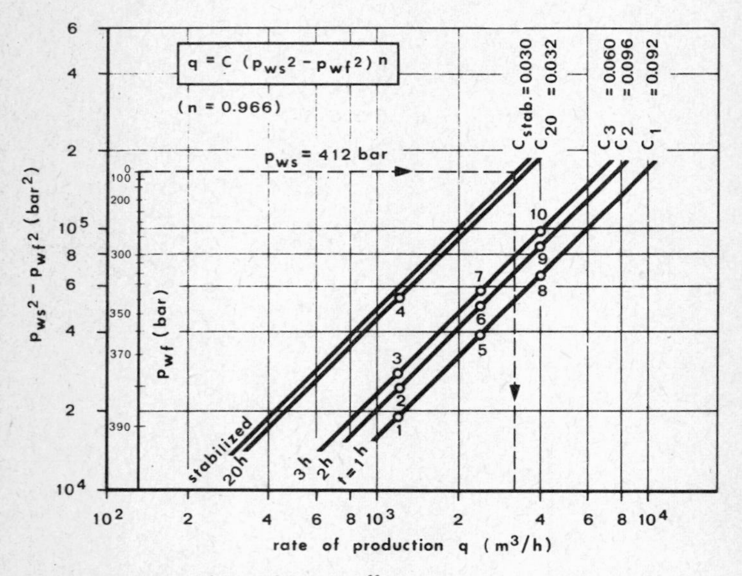

Fig. 122 Isochronal Test in gas wells.

The well is produced for a certain period t and then shut in for the same time period. Then the well is produced again for a certain period at a constant rate which, however, differs from the first one etc. The flowing pressures (as $p_{ws}^2 - p_{wf}^2$) vs. the related production rate q are plotted on a logarithmic scale as in Fig. 121. The results are parallel straight lines the inclination of which is representative of the "well's behaviour": if n would be 1 (inclination and tg $\alpha = 45°$), the conditions would conform exactly to Darcy's Law. Deviation of this "ideal" 45 ° – value point to factors which influence the flow of gas (i.g. viscosity which is dependent on the pressure, furthermore the pore geometry determining the degree of turbulence of the flow; *Houpeurt* 1959[8]). The position of the straight lines, as described by C, depends on the time t : 1, 2, 3 and 20 hours in the above example. The longer the test period the more stabilized the flow conditions are. See Appendix M

Appendix M, which shows the proposed procedure for measurements and subsequent evaluation of an isochronal test.
Instead of pressures measured downhole in the wellbore, well head pressures, too, may be used to determine the capacity (Cullender[3]). Naturally in this case no liquid should be present in the hole, as this would cause erroneous pressure values to be used.
The capacity diagram obtained applies to the very moment when it is recorded. When in the course of production, reservoir pressure declines, a change of C is to be expected requiring the measurement of a new capacity diagram.
Gas wells producing from reservoirs of low permeability require some time to remove the liquid that has penetrated into the reservoir rock in the vicinity of the well (mud filtrate, acid, water etc.). In such cases the capacity test yields lower values than after prolonged production. *Leicht*[12] has developed a semi-empirical method for the correction of results obtained when the well has not yet been cleaned up. This method is based on the fact that these impurities reduce the effective permeability in the vicinity of the well. The effective permeabilities computed from pressure build-up measurements (refer to P. 129) are used to determine a corrected C constant.

4.1.3 Pressure Build-up Tests

More knowledge about the reservoir can be obtained in the first instance by the pressure build-up curves. Demonstrated for the first time by *Horner*[7] at the Third World Petroleum Congress of 1951, they constitute an important step forward in determining the important characteristics of oil and gas reservoirs.

4.1.3.1 General

After a producing well has been shut in, the pressure in the hole (usually recorded by a recording pressure bomb mounted opposite the pay) will rise continuously. From the rate of this pressure increase, valuable information on several reservoir parameters may be derived as will be shown further below. Fig. 123 shows pressure values recorded in a well of a shallow oil-field both during production and after shut-in. A detailed description is given in Appendix N.
Usually, when the pressure bomb is pulled the shut-in pressure has not yet reached its final value. In reservoirs of low permeability it often takes days, sometimes weeks for this pressure build-up to be completed. *Horner* developed a method to determine in an economic period of shut-in time, the static pressure which the bomb would have recorded after infinite shut-in time in an ideal reservoir. In other words, the pressure prevailing at the "border" of an infinite reservoir, i.e. the pressure p_i (i stands for infinite or initial), can be determined.

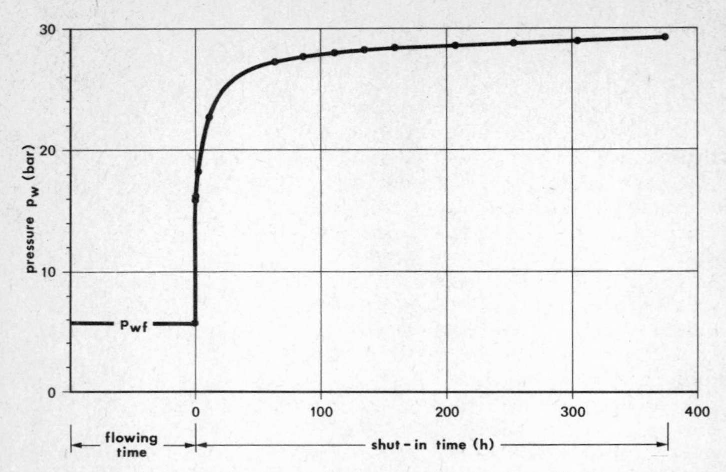

Fig. 123 Pressure measurements in a shallow oil well of a German field. Depth of measurement (and depth of the pay horizon) is 301 m. The well was flowing at a rate of 4 m^3/day with a flowing pressure (p_{wf}) of 5.6 bar. The well was then shut in (time Zero hours). The pressure "builds up", quickly at first, then more and more slowly. After a shut-in time (Δt) of almost 380 hours the pressure bomb was pulled. At this time the shut-in pressure is around 29 bar and it is still rising. For more details see Appendix N.

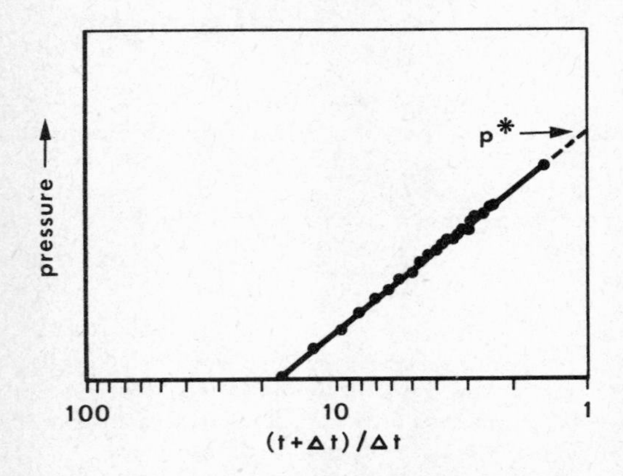

Fig. 124 Pressure build-up in an "ideal reservoir" vs. ln $(t + \Delta t)/\Delta t$. Extrapolating of the straight line against ln $(t + \Delta t)/\Delta t = 1$ yields the real shut-in pressure designated by p*. (After *Matthews* and *Russell*[13])

The law governing pressure build-up in a shut-in well is expressed in a formula, which in practical units reads as follows:

$$p_{w\Delta t} = p_i - \frac{21.91 \cdot q \cdot \mu}{B_o \cdot k \cdot h} \log \frac{tp + \Delta t}{\Delta t}$$

where

$p\Delta_t$ = pressure in the hole after shut in time Δt (bar).
p_i = original reservoir pressure in an infinite reservoir (bar).
q = production rate of the well (m^3/d).
μ = viscosity of the liquid under reservoir conditions (cP).
B_o = formation volume factor.
k = permeability of the reservoir rock (md).
h = productive thickness (m).
tp = production time of the well from the start of production to the time it was shut-in (h).
Δt = shut-in time (h).
A graphical presentation of the above equation is shown in Fig. 124 plotting p_w vs.

$$\log \frac{tp + \Delta t}{\Delta t}$$

The extrapolation of the straight-line part of the curve yields the static pressure we would obtain in a reservoir of infinite extent after an infinite shut-in time. This extrapolated static pressure is generally designated by the symbol p^*.

$p_{\Delta t}$ plotted against $\log \frac{tp + \Delta t}{\Delta t}$ yields a straight line, the inclination of which is defined by $\tan. \alpha = \frac{21.91 \cdot q \cdot \mu}{B_o \cdot k \cdot h}$. The permeability k can be computed from this equation, as all other values are known or measurable. In practice this computation is simple, as tan. α equals the pressure difference between two logarithmic cycles. In the case of the pressure build-up of Fig. 125 this is 26 bar per log cycle or:

$$\tan. \alpha = m \text{ (bar/log cycle)}$$

$$k = \frac{21.9 \cdot \mu \cdot q \cdot B_o}{h \cdot m} \text{ (md)}$$

In Fig. 125 which is described in more detail in Appendix N, k is calculated to be 19.6 md.
In an "ideal case" influx from the reservoir into the well would stop as soon as the well is shut in (which, as a rule, is done at the surface). This, however, is not the case. When the well is shut in,

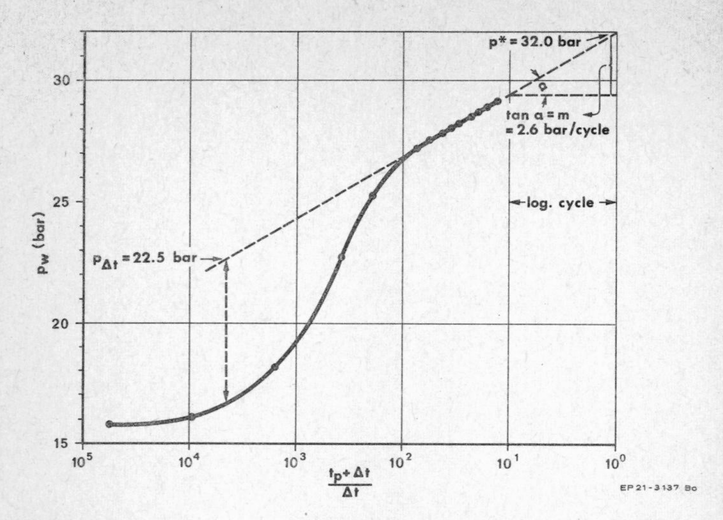

Fig. 125 *Horner*-interpretation of the pressure build-up curve shown in Fig 123. When plotting the shut-in pressure $p_w\Delta_t$ (where Δt is the time, in hours, since closing the well) vs. log $\frac{tp + \Delta t}{\Delta t}$ (where tp is the time since start of the production or since the last shutting in of the well) a part of the pressure build-up results in a straight line. When this line is extrapolated against $\Delta t = \infty$ (in other words: log $\frac{tp + \Delta t}{\Delta t} = 1$) the reservoir pressure (maximum shut-in pressure) is reached. This extrapolated static pressure is generally designated by the symbol p*. The inclination of this straight line is α. Tan α = m = pressure per one logarithmic cycle is 2.6 bar in this well. The permeability of the pay horizon may then be calculated. The value of $p_{\Delta 1}$ is a function of after-production plus skin effects

influx into the well continues, and the term "after-production" or "after-flow" designates that part of production still flowing into tubing and casing at a gradually decreasing rate after the well has been shut in. This after-production causes pressure build-up curves to deviate from the straight line (refer to Fig. 125 and Fig. 127). Another factor contributes to deviation of the pressure build-up curve from a straight line at the start of the shut-in time. This is the so called "Skin-Effect". In form and size, the funnel-shaped "depression" around a producing well depends on the production rate, the formation's resistance to flow, the viscosity of the oil and an additional resistance to flow restricted to a (more or less thick) zone directly adjacent to the wall of the hole. This additional resistance to flow may result from the way the well has been completed (perforation, slotted pipe, open hole etc.) or from the blocking of flow channels caused by infiltration of mud, cement or water, by formation of emulsions from swelling of clays etc. The pressure drop caused by this additional resistance to flow was called "Skin Effect" (refer to Fig. 126) by *Van Everdingen* (1953)[5].

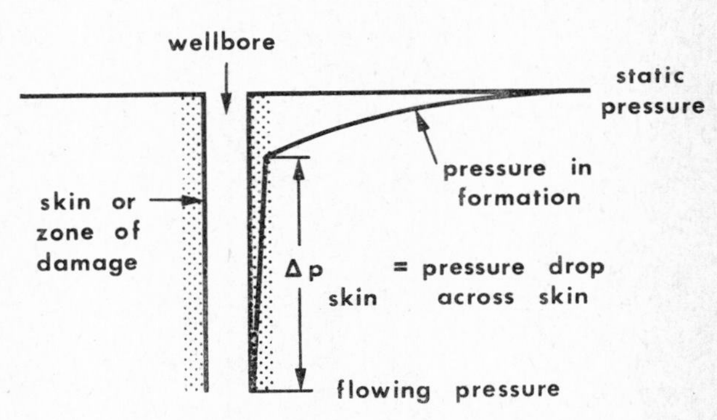

Fig. 126 Pressure distribution in a reservoir with a skin. The additional resistance to flow is caused by blocked flow channels. (After *Hurst*[5])

The amount of this initial deviation from the straight line curve may be expressed by the term $p_{\Delta t}$, which is the difference in pressure between the actual measurement and the backward extrapolation of the straight-line part of the curve measured at Δt = one hour. (See Fig. 125.) It is an expression of the effects of after-production + additional flow resistance.
To characterize the permeability of the reservoir rock directly around the well, the term "Skin Factor" (S) was introduced. This factor can be computed by the following formula:

$$S = 1.151 \frac{p^* - p_{wf}}{m} - \log \frac{7.826 \cdot t_c \cdot k}{\phi \cdot \mu \cdot c_0 \cdot r_{w2}}$$

In Fig. 125 which is described in detail in Appendix N the Skin Factor S amounts to + 3.02.
One should note that the Skin Factor gives a qualitative indication merely for an increased or reduced flow resistance in the immediate vicinity of the well bore. A positive Skin Factor of + 3 to + 5 (sometimes up to +10) is customary for oil wells.

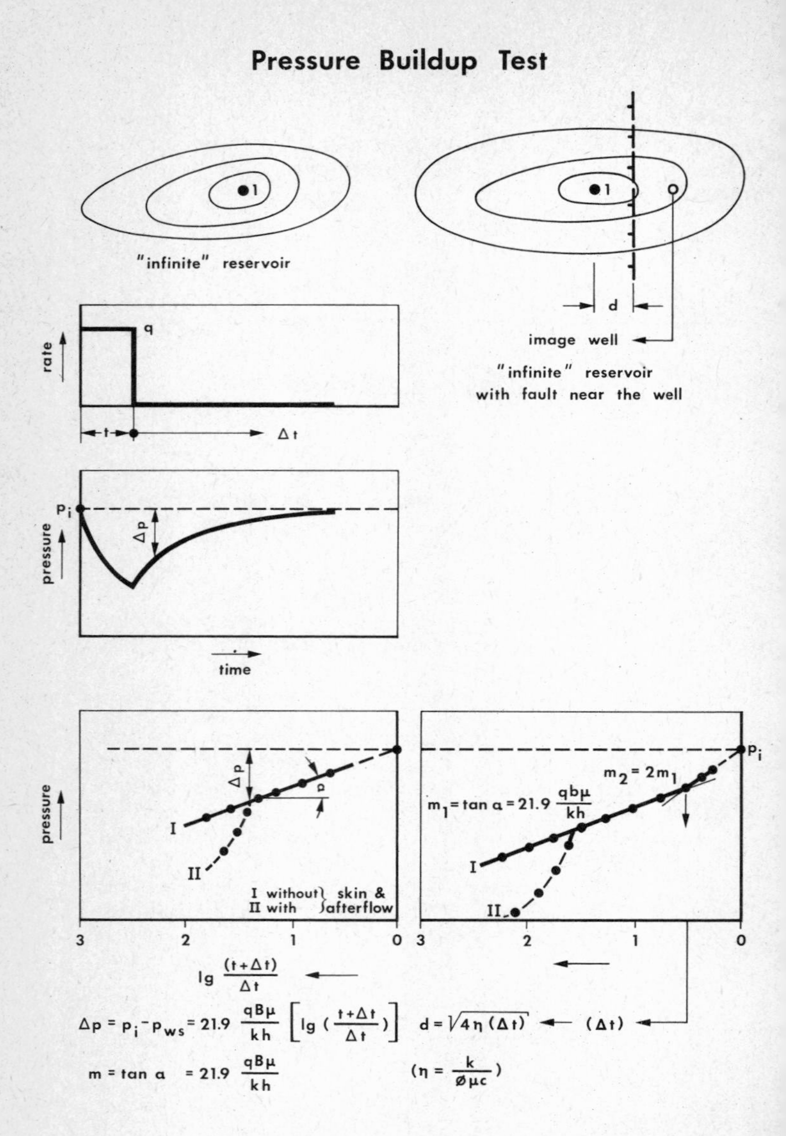

Fig. 127 Pressure Build-up Test

In gas wells it may be even higher (but in most cases merely apparently caused by turbulent flow). A negative Skin Factor means that flow conditions near the wall are better than in the reservoir at large, e.g. as a result of succesful acidizing or fracturing jobs.

4.1.3.2 Special cases

"Ideal" pressure build-up curves as described above and represented in Fig. 125 will result from producing wells in infinite i.e. unlimited reservoirs. Though there is no such reservoir in nature, some behave in this manner if the production period of the well is not extended over too long.

Usually, however, the reservoir is limited, and/or the well's drainage area is finite. This will be especially so when, the "pressure funnel" generated by production, reaches the border of the reservoir, or when a well is immediately surrounded by several adjoining wells. In such cases the build-up pressure will no longer be p* resulting from an extrapolation of the straight part of the pressure build-up curve but p_{ws} as indicated in the final part of the pressure build-up curve (see Fig. 128).

When, in an otherwise infinite reservoir, a tight fault occurs at a distance "d" from the producing well this fault shows up characteristically in the pressure build-up curve. In logarithmic representation the curve develops a break, whilst the inclination of the first part is $\frac{q\,\mu}{4\,kh}$ that of the second part is $\frac{q\,\mu}{2\,kh}$. This is shown in more detail by Fig. 129. In certain cases the distance

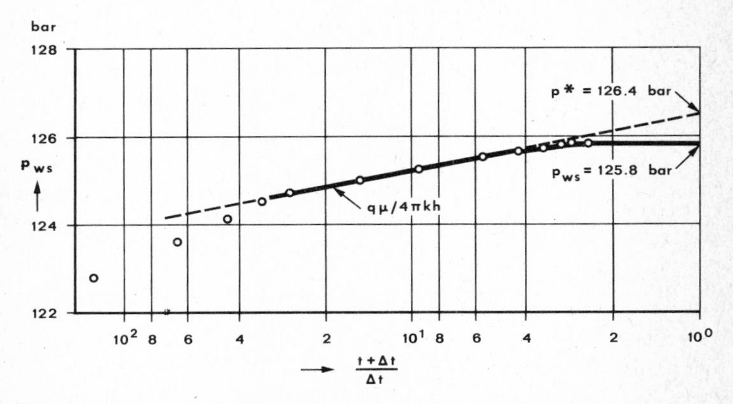

Fig. 128 Pressure build-up curve in a limited reservoir. An extrapolation of the straight-line part of the curve yields a p*-value of 126.4 bar while a break in the final part of the curve leads to a shut-in pressure p_{ws} of 125.8 bar (Steimbke-Ost Nr. 4)

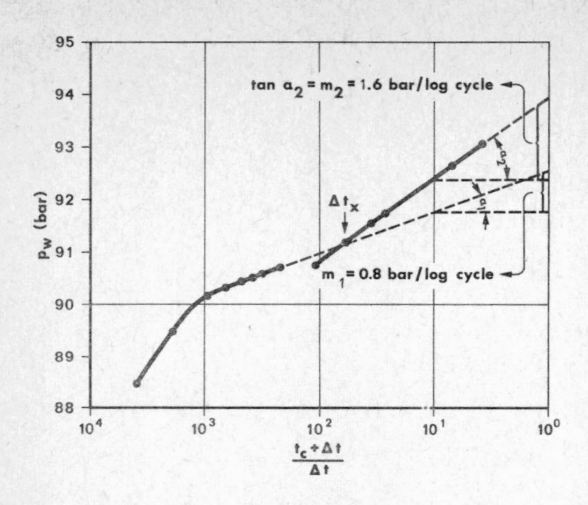

Fig. 129 Build-up pressure curve of the well with a nearby fault. Tan α_1 (= m_1) equals half of tan α_2 (= m_2). Plotted on logarithmic paper

"d" to the fault (not the direction) can be computed using a — Ei(— x —) function solution.

Pressure build-up curves also deviate from the usual form when the reservoir rock is composed of layers of different permeability. In this case, pressure build-up will be completed first of all in the more permeable layer whilst fluid will keep flowing from the less permeable layer (where pressure is then higher) into the more permeable layer. The result is a pressure build-up curve as shown in Fig. 130. The same type of curve will result when the reservoir is fissured.

In wells with higher gas/oil ratio the pressure build-up curve often shows a pronounced hump which, in all probability, is caused by a rise of gas bubbles in the production string (see Fig. 131).

If a well which was originally shut in is opened, one arrives at pressure decline developments. Pressure decline curves may be handled in the same way as pressure build-up curves. In wells with higher gas/oil ratio, the first part of the curve will be a curved one due to the liberation of gas in the production string and by the removal of the oil in the annular space. The following section will be linear (in logarithmic representation) and permit the kh product to be computed similarly to pressure build-up curves. Finally, the above arithmetical and graphical evaluations may also be applied to pressure decline curves of injection wells when they are shut in.

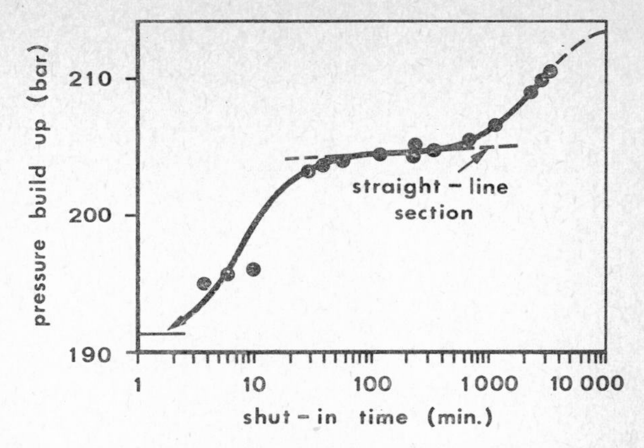

Fig. 130 Pressure build-up in a two-layer reservoir with different permeabilities (after *Matthews, C. S., D. G. Russell*[13]). After a levelling-off (straight-line section) the pressure rises again and then levels off finally

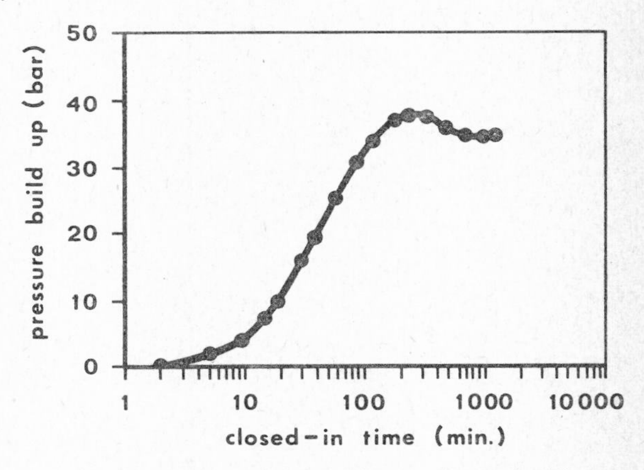

Fig. 131 Certain wells exhibit a peculiarity during build-up – that of "humping" which is due to a rise of gas in the tubing after closing – in. (After *Matthews, C. S., D. G. Russell*[13])

If, for instance, in a gas-field the flow is of a semi-steady state and if the drainage area boundaries equal the reservoir limit, then the "gas in place" can be calculated using pressure build-up curves[4] according to the formula

$$G = \frac{0.1752\ p_{ws}\ q\ \Delta t}{z\ m}$$

in which

G = Gas "in place" (m^3)
z = Super-compressibility factor of the gas

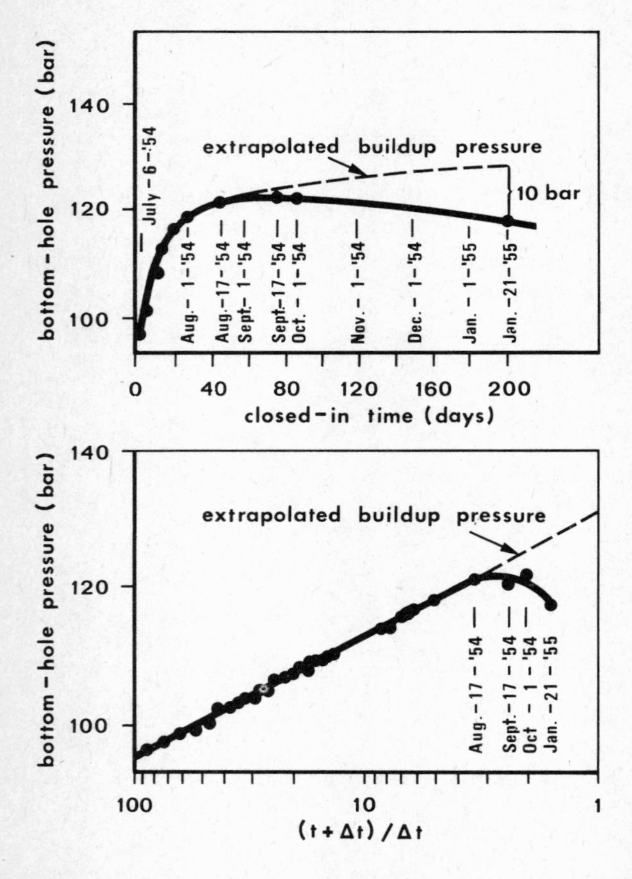

Fig. 132 Interference test in a low permeability reservoir (after *Matthews* et al.[13]). The dotted line in the upper figure (called "extrapolated build-up pressure") was obtained by extrapolating the linear portion of the log plot shown in the lower figure

4.1.4 Interference and Pulse Testing[6]

Other important types of pressure build-up measurements are the so-called interference tests. These are measurements of the pressure in a well which is exposed to the influence of a neighboring producer or injector. It may take months for such an interference to become determinable (Fig. 132). With such measurements it is also possible to determine the reservoir rock's

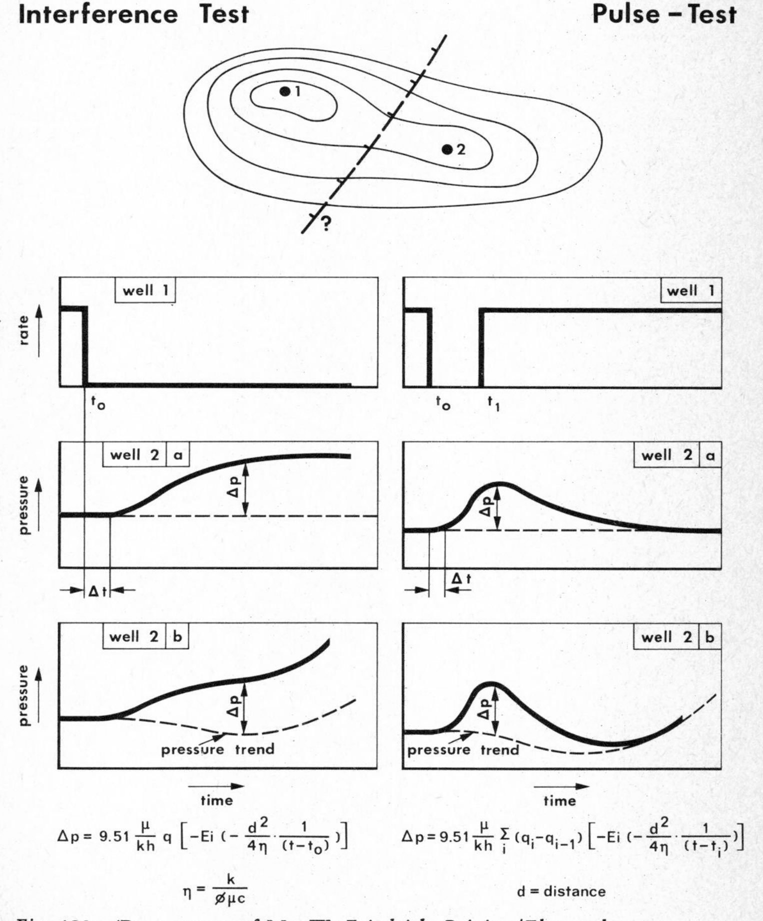

Fig. 133 (By courtesy of Mr. *W. Friedrich, Brigitta/Elwerath,* Hannover). See page 138

Left-hand side Fig. 133 (page 137):

The "interference test" is an almost classical method of investigating a reservoir between two wells. Well 1 is closed in, and pressure changes are observed in the adjacent well 2. If there is no fault between the two wells, a pressure performance roughly corresponding to the diagram in the middle (well 2 a = case without pressure trend) may be expected. If there is a fault in between, the pressure rise will be either reduced or non-existent. Difficulties occur when the general pressure does not remain constant (as shown by the dotted line): in the lower diagram (well 2 b = case with pressure trend), a general pressure trend in the vicinity of the measuring well has been assumed (dotted line). In such a case it is very difficult, if not impossible, to determine an interference pressure rise. This difficulty may be overcome by the pulse test method.

Right-hand side Fig. 133:

Problems with interference tests, as described above, can be overcome if well 1 is closed only for a certain period of time and then reopened. This is called a pulse test. Well 1 is the pulsing, well 2 the responding well. In well 2 a (case without pressure trend), a pressure rise is observed, followed by a pressure decline. Shape and amplitude of the measurable pressure pulse in well 2 a depend on the reservoir parameters in the influence area of the pulse test and, of course, on the length of the closing time. It is now possible to conclude from the general form of the pressure curve the existence of a general pressure trend in the vicinity of the well (compare lower diagram of well 2 b which represents the case with an assumed pressure trend).

porosity, a parameter which pressure build-up measurements not exposed to interference fail to consider.

Special measurements are necessary in the borehole to record the small pressure pulses of interference and pulse testing. This is shown in Fig. 134.

After the pioneer work of *Johnson* et al.[9] and *Brigham*[2], the pulse test became an important diagnostic technique of the reservoir engineer. Its application was advanced as high-precision instruments became available for pressure measurement. As pressure pulses during pulse testing are usually relatively small — depending on the parameters of the reservoir —, their identification and quantitative determination require exact measurements to be taken. For practical reasons, the "transmitting" well is pulsed at least twice, whereby the total measuring time is reduced considerably. Moreover, when interpreting the results one does not use the absolute pressure differential Δp, but the tangential pressure differential Δp^* (see Figure 134).

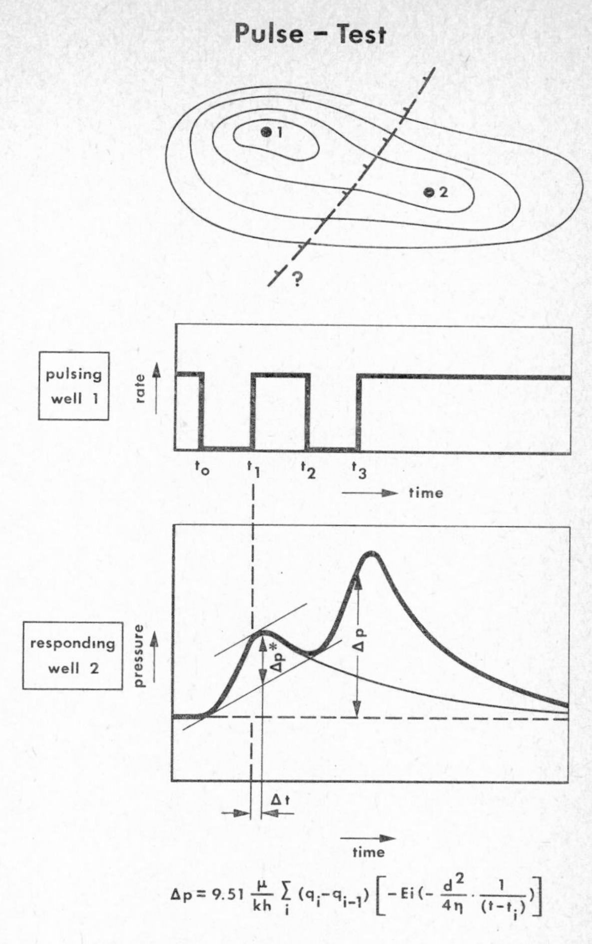

$$\Delta p = 9.51 \frac{\mu}{kh} \sum_i (q_i - q_{i-1}) \left[-Ei\left(-\frac{d^2}{4\eta} \cdot \frac{1}{(t-t_i)}\right)\right]$$

Fig. 134 (By courtesy of Mr. *W. Friedrich*)

In pulse test calculations in faulted areas, each producing well has to be "mirrored" symmetrically to the fault by a well with an equal production history. The calculations then use the principle of superimposition. An analysis of the results and also the planning of pressure pulse measurements may be done using the elegant methods of *Brigham* (loc. cit.).

It should also be mentioned that a fault may be detected by pulse testing when it runs parallel to the "transmitting" and "receiving" well. Its presence is then indicated by a steeper rise of the pressure pulse.

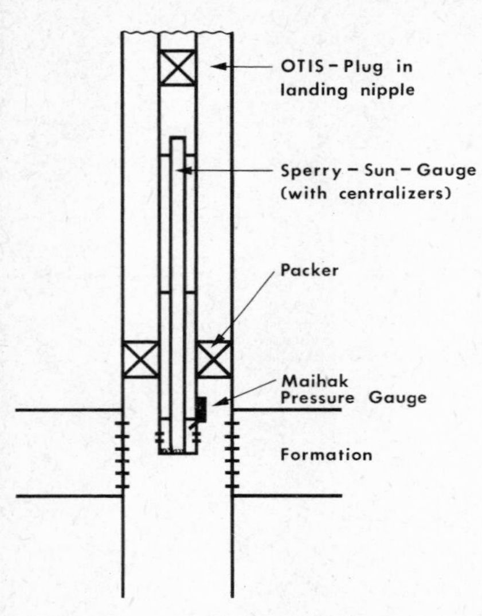

Fig. 135 As the pressure pulses during interference and pulse testing are generally very small, special measures are necessary for a reliable detection. On occasion of such a test run in a German oil-field the above wellbore installation was used with two highly sensitive pressure measurement instruments

4.1.5 Flow tests

A drawback of pressure build-up curves is due to the well is being closed in and – at least when the reservoir rock is of low permeability – it must be kept closed for a considerable time. In addition, an analysis of the results is often problematic. To avoid these disadvantages, *Russell*[14] published a method offering the same results as the pressure build-up curves (determination of static pressure, of permeability and of the skin effect). The method consists in a change of the well's production rate and an evaluation of the pressure response. However, subsequent to the change of the production rate an unsteady flow will prevail until conditions have stabilized (unsteady state, transient conditions, refer to page 77).

This method of applying pressure conditions to the new production rate is shown in Fig. 136. As in the case of pressure build-up curves, the behaviour as represented in Section A is used to determine static pressure, permeability and the skin effect. In limited reservoirs, or when neighboring wells interfere, effects similar to those of pressure build-up curves show up. (In Fig. 136 Section A has been over-extended with regard to t for greater clarity. Actually, stabilization sets in after a few hours or days.) More details and an example from a German oil-field are given in Appendix O. As a rule, the production rate is reduced, as a reduction leads to a quicker stabilization of production and pressure conditions. How-

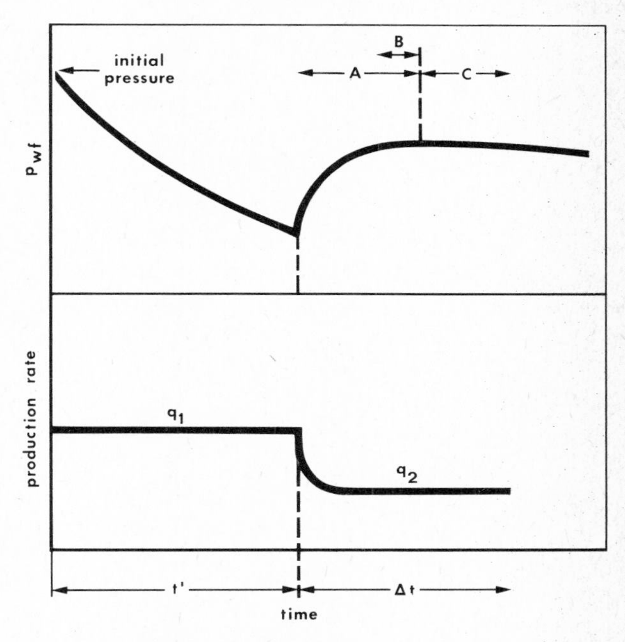

Fig. 136 Diagrammatic representation of the pressure change caused by a change of the production rate (after *Russell*[14]). As a consequence of the production rate q_1, the pressure on the bottom of the hole has in the course of the time t′ dropped down to a value measured immediately prior to the change of the rate of production. After changing the rate to q_2 it will take a certain time to stabilize the new rate (Section A which includes certain boundary and interference effects marked by B). Flowing pressure p_{wf} will change accordingly. After some time, the pressure will have stabilized too (Section C). The time elapsed since the change of the production rate is defined as Δt

ever, the method is also applicable when the production rate is increased. The method offers the same possibilities of application as the pressure build-up method. It may also be used in injection wells.

4.1.5.1 Reservoir limit tests

When a wildcat has discovered a new field, the first question concerns the size of the reservoir. In addition to the possibilities discussed in Section 5.1. there are ways and means of drawing certain conclusions with respect to the extent of the reservoir from initial production and pressure performance.
One way of getting information on the amount of oil in place from pressure build-up curves has been shown on page 135. Basic conditions[5]:

a) The well must have produced long enough for its drainage radius to have reached the borders of the reservoir.
b) Shut-in time for the purposes of pressure build-up measurement must extend long enough for the change of the pressure build-up to be recognized.
c) Computations have to assume a homogeneous reservoir (of constant formation thickness, constant porosity and constant permeability).
The formula reads:

$$-\mathrm{Ei}\left(-\frac{2870\,a^2\,\Phi\,\mu c}{t_c\,k}\right) = 2.3 \log \frac{t_c + \Delta t}{\Delta t}$$

in which a = distance to the "border" of the field. – Ei (–x) is taken from mathematical "tables of higher functions".

The procedure is simple and convenient, but often it is very hard to derive properly the dip angles in the pressure build-up diagram. Theory requires the second (the latter) straightline part of the pressure build-up diagram to show double the inclination of the first one.
Evaluations from flowing pressures are possible with the following basic conditions:

a) The well must have produced long enough for its drainage radius to have reached the reservoir borders.
b) Constant flow rate must be maintained during the test.
c) Homogeneous reservoir properties are assumed.

P. Jones (1962)[11] evaluates flow pressure observations made at a constant rate of production. Without going into details it should be mentioned that in gas reservoirs the availability of data often enables interesting conclusions to be drawn with respect to reservoir size and form, to the presence of a gas/water contact, to the angle between two nearby fault lines and to the number of nearby faults etc.

L. G. Jones (1963)[10] also evaluates flow pressure observations made at a constant rate of production. He establishes the relation between cumulative gas production G_p and the quotient

$$\frac{p_w^2}{G_p}$$

in which p_w^2 is the difference between the squares of two subsequent flowing pressure observations, and G_p is the volume of gas produced between these two flowing pressure observations. In the course of the early, unsteady part of the test performance the values of p_w^2/G_p yield a straight line when plotted against log G_p while later, under steady conditions, the curve of the p_w^2/G_p values forms a straight line when plotted against G_p. These diagrams then permit the determination of the gas in place.

4.2 Production control measurements

These measurements are used to check the influx conditions of producing oil and gas wells. Among these procedures those practised by Messrs. *Schlumberger* will be discussed mainly at this place as they are most commonly used in Germany. Beyond the devices developed for *Schlumberger* procedures a variety of others have been developed; but it would grossly exceed the scope of the present chapter to mention all of them. Brief reference should be made, however, to the oil/water contact measuring instruments developed by *Leutert* (*Bögelmann*[1]). They register the position of the oil level and of the oil/water contact in shut-in wells and some of them may be combined with temperature and pressure measuring instruments. Oil/water contact measurement by these instruments is based on the difference between the resistivities of the two media.

The production control measurement procedures developed by Schlumberger may be classified into two main categories:

1. Instruments for continuous measurement which may be operated in combination:
Continuous Flowmeter
Gradiomanometer
High Resolution Thermometer
Manometer
Through Tubing Caliper

2. Instruments for stationary measurement which may only partly be operated in combination:
Packer Flowmeter
Densimeter
Water-cut-meter (Slugometer)
Fluid Sampler

The "continuous flowmeter" diagrammatically represented in Fig. 137 is operated at constant speed against production flow.
The "gradiomanometer" measures the pressure difference between two diaphragms arranged at a distance of two feet; it transmits this difference electrically to the surface. The device detects the entry of large quantities of a lighter medium into a heavier one.

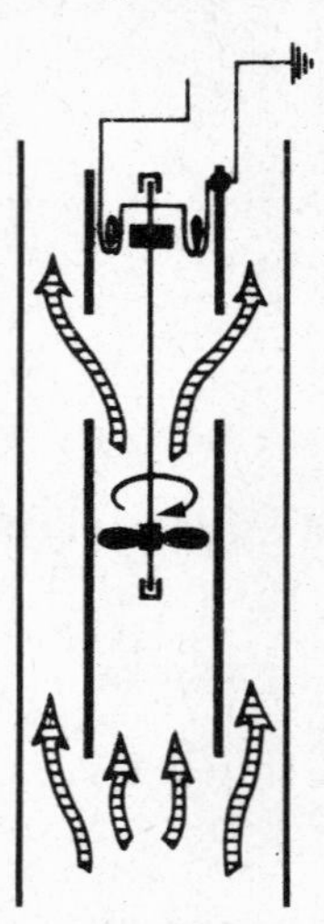

Fig. 137 Continuous Flowmeter. The flow causes a propeller to rotate, an attached magnet induces an alternating current in a coil. The frequency of this current depends on the rotational propeller speed and this on the flow velocity. Recording is done at the surface and permits a determination of the production rate

The "high resolution thermometer" records temperatures with a resolving power of 0.2 °C. Measurement is made by an electrical resistance wire which is sensitive to temperature. As it permits a recognition of the adiabatic cooling effect it is mainly used to detect gas penetration points in a well. However, penetrations of other media are detected as well, as these necessarily modify the natural temperature gradient prevailing below the point of penetration. Even flow processes on the outer casing side may be detected, so this device constitutes a valuable accessory to the two devices already described.
The "manometer" consists of a temperature-compensated bourdon tube. It is connected to a potentiometer and continuously registers pressure changes in the well.

The "through tubing caliper" is a three-armed caliper with a measuring range of 18″–2″ at a minimum tubing diameter of 1 11/16″.
The instruments for stationary measurement are mainly used in wells of low influx rates.
The measuring principle underlying the "packer flowmeter" is the same as that of the continuous flowmeter. In this case the absolute velocity of the flowing medium is measured, as the instrument itself is stationary when making the measurement. An inflatable packer sealing the space between the instrument and the wall of the borehole forces the total flow to pass the device.
Two other instruments can be used in combination with the packer flowmeter. The "densimeter" measures the density of the fluid in the hole. The density of a vibrating body affects the frequency of oscillation which is recorded electrically at the surface. The "slugometer" measuring the dielectric coefficient of the flowing medium is an excellent means of locating points of water influx. The "fluid sampler" takes liquid samples under reservoir conditions at any desired depth. A two-chamber system prevents pressures losses during sampling and transportation.

Bibliography

[1] *Bögelmann, K.:* Zwei neue Meßgeräte für die Fördertechnik. Erdöl-zschr. 80, Wien 1964.

[2] *Brigham:* I. Petr. Techn., May 1970.

[3] *Cullender, M. H.:* The Isochronal Performance Method of Determining the Flow Characteristics of Gas Wells. Trans. AIME, Techn. Pap., 4120, New York 1955.

[4] *Cunningham* und *Nelson*, J. Petr. Techn. Juli 1967, p. 859.

[5] *Everdingen, A. F. van, W. Hurst:* The Application of the Laplace Transformations to Flow Problems in Reservoirs. J. Petrol Techn. London AIME, Petrol Trans., 168, New York 1949.

[6] *Friedrich, W.:* Druckaufbau – Interferenztest – Pulstest – klassische und moderne Werkzeuge des Lagerstätteningenieurs im praktischen Einsatz. Erdöl-Erdgas-Zschr. April 1976.

[7] *Horner, D. R.:* Pressure Build-up in Wells. Proc. Third World Petrol. Cong., Sect. II, The Hague 1951.

[8] *Houpeurt, A.:* Rev. Inst. Franc. Pétrol, 11 und 12, Paris 1959.

[9] *Johnson et al.:* I. Petr. Techn., Dez. 1964.

[10] *Jones, L. G.:* Reservoir Reserve Tests. J. Petro Techn. 15, London, March 1963.

[11] *Jones, P.:* Reservoir Limit Test on Gas Wells. J. Petr. Techn. 14, London, June 1962.

[12] *Leicht, H.:* Korrektur der Kapazitätsgleichung bei nicht freigeförderten Gasbohrungen. Erdöl-Zschr. 79 (1963).

[13] *Matthews, C. S., D. G. Russell:* Pressure Buildup and Flow Tests in Wells. New York 1967.

[14] *Russell, D. G.:* Determination of formation characteristics from two-rate flow tests. J. Petr. Techn. 15, London, Dec. 1963.

5 The development of oil- and gas-fields

The development of a newly discovered field is a technique which calls for a great deal of experience and for the application of practically all factors discussed in the previous chapters. The target is clearly defined: to recover as much oil as possible at as low a cost as possible. Much depends on the drilling plan, i.e. at which place (distance) and, in which period of time, to drill how many wells. All parameters of the reservoir and its content affect the answer to this question. Additionally, a clear geological and geophysical picture of the field is needed to solve this task. Furthermore it is obvious that the technical installations (from the bottom of the well to the delivery point at the border of the field – the job of the mechanical engineer) are to a great extent affected by the development planning of the petroleum engineer. Finally both departments have to consider all their proposals in the light of economic considerations which forces engineers and geologists to become familiar with the basic economic principles and the related yardsticks of investments, costs and surplus.

The problems which govern the development of an oil-field are different from those of a gas-field: while an oil-field can be developed from a purely technical and economic viewpoint, the development of a gas-field is additionally influenced by the wishes and requirements of the consumers. It is a product of engineering plus sales requirements.

In each case the first task after the testing of a discovery well is to estimate the size of the new discovery. At this stage the petroleum engineer will be grateful for any information, however small. Experience has shown that, therefore, in the discovery well it always pays to do "a bit too much" as far as the following measures are concerned: coring of the pay horizon, intensive logging, fluid-sampling and analysis, pressure measurement during the testing period and all kinds of lab-work.

5.1 Estimation of reserves

It goes without saying that knowledge of the available reserves is indispensible in planning all activities and in determining the economics. This means that the first two things to do after the discovery of a new field are to find out the production capacity of the well and to estimate the reserves.

Obviously, the sooner such information on recoverable reserves is available and the preciser it can be, the greater is its value. But such an early availability and, in particular, such early exactitude is not easy to obtain; in other words, the later such computations are made, the more exact and the less interesting they will be.

When the stage of production is used as a criterion, the methods of reserve computation may be classified as follows:

1. Methods applicable at an early stage; in particular the volumetric method. This is practicable as soon as the borders of the field are at least roughly known. Its weakness lies in the fact that both the border and the recovery factor can be more accurately determined only when more experience is available.
2. Methods applicable after a certain period of production, e.g. the computation of original reserves with the help of material balance calculations. As already shown, such calculations are possible, however, when the reservoir pressure has declined somewhat.
3. Methods applicable after an extended period of production. These methods generally use decline curves. Although the "forecast" obtained by them is very accurate, they become less and less interesting as the field grows older.

Table 2 Reserve calculations

method	time	gives	comments
A. static			
1. volumetric	early	oil/gas "in place"	becomes accurate only later (when more data are available)
B. dynamic			
2. material balance	later (after producing some 10 % of the reserves)	oil/gas "in place"	requires periodic repetition
3. decline curves	late	recoverable reserves	fairly accurate
4. pressure build-up curves	early	gas "in place"	only if pressure influence reaches reservoir limit
5. reservoir limit rest	early	rough indication of pore volume, oil/gas "in place", reservoir radius	useful as additional information in case of prolonged production at ± constant rates

Another classification of methods is the following:
1. Methods indicating the "oil originally in place" while the recoverable part of these reserves has to be estimated. In this category belong the volumetric and the material balance methods.

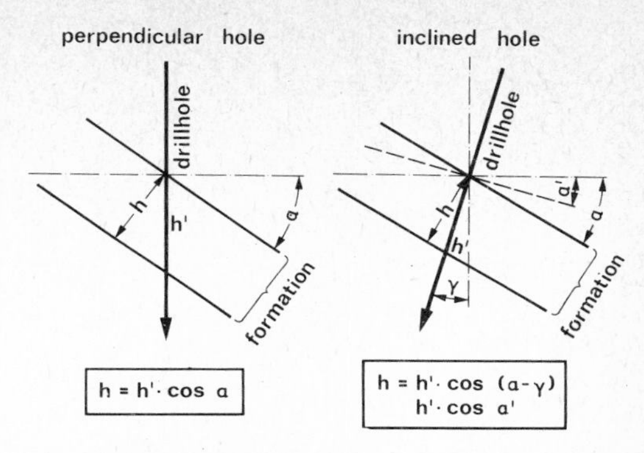

Fig. 138 When determining the true thickness h of the pay horizon, the dip of the formation and the inclination of the drill hole must be taken into account

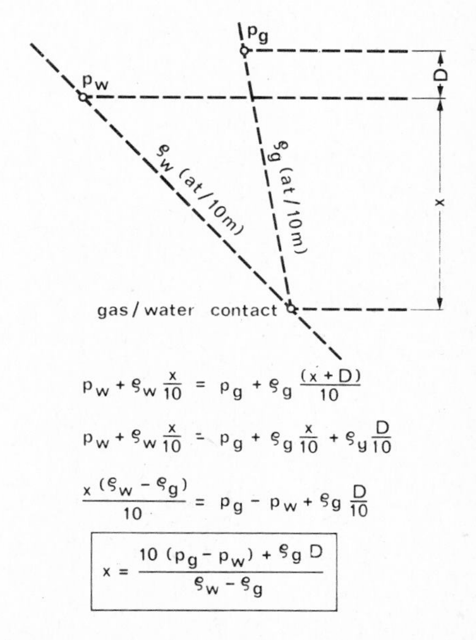

$$p_w + \varrho_w \frac{x}{10} = p_g + \varrho_g \frac{(x + D)}{10}$$

$$p_w + \varrho_w \frac{x}{10} = p_g + \varrho_g \frac{x}{10} + \varrho_g \frac{D}{10}$$

$$\frac{x\,(\varrho_w - \varrho_g)}{10} = p_g - p_w + \varrho_g \frac{D}{10}$$

$$x = \frac{10\,(p_g - p_w) + \varrho_g D}{\varrho_w - \varrho_g}$$

Fig. 139 If the pressure in the gas column of a gas well (p_g) and that in the water-bearing part of the structure (p_w) is known, the depth of the gas/water contact (and with it possibly the productive area) may be calculated

2. Methods indicating the recoverable part of reserves while the original reserves have to be estimated. One of these methods is that of production decline curves.
In the following, the methods are classified as static (volumetric) and dynamic (refer to table 2).

5.1.1 Volumetric method

The volumetric method requires the exactest possible data on
a) the thickness of the reservoir rock,
b) its extension,
c) its porosity and
d) its saturation.

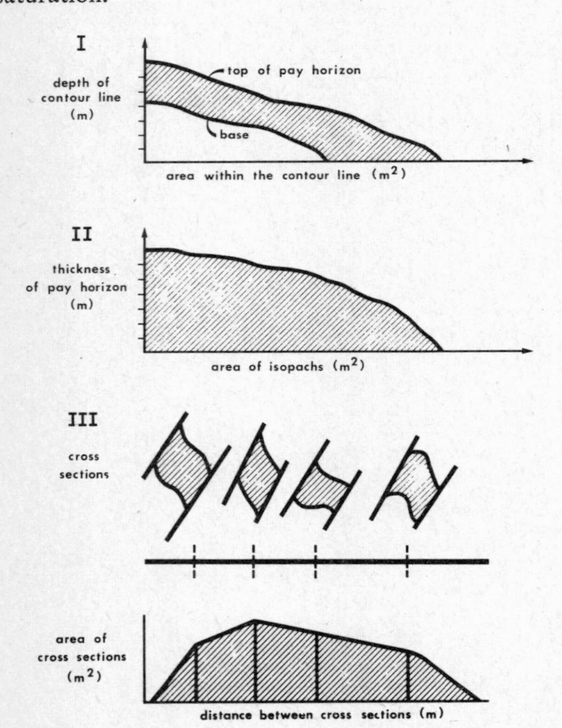

Fig. 140 Three methods for volumetric determination of pay horizon. Method I is based on contour maps of top and base of the pay. Method II uses isopach maps of the pay. Method III is applied to tectonically complicated structures. In each case, the shaded area must be determined by planimetry.

By a multiplication of these 4 factors we then get the original oil reserves in the field under reservoir conditions.
The thickness of the pay horizon is usually obtained from downhole measurements (SP, resistance). As only the net thickness – the producible part of the reservoir rock – is measured, additional

Table 3 Calculation of reserves by the volumetric method

$$N, G = \frac{A \times h \times \phi \times (1 - S_w)}{B}$$

N = oil "in place", m^3

G = gas "in place", Nm^3

A = productive area, m^2

h = net thickness, m

ϕ = porosity

S_{wi} = connate water (irreducable)

B = formation volume factor

for gas: $$\frac{1}{B} = \frac{p_i}{p_{atm}} \frac{T_{atm}}{T_r} \frac{1}{z} = E$$

p_i = initial reservoir pressure, bar

p_{atm} = atmospheric pressure, bar

T_r = initial reservoir temperature, °Kelvin

T_{atm} = 0 or 15 °C $\begin{matrix} 273.16 + 0\ °C\ (= Nm^3\ 0\ °C) \\ 273.16 + 15\ °C\ (= m^3_s\ 15\ °C) \end{matrix}$

z = super compressibility factor

Table 4 Volumetric reserve calculations

thickness (h)

relevant terms:

1. true and apparent thickness (dip of formation and inclination of drill hole)
2. gross thickness
   a) by coring
   b) from cuttings
   c) from rate of penetration logs
   d) from geophysical well surveys
3. net thickness
   (only porous and permeable portions)
   determined as under 2 a) and 2 d)

information is required from micrologs and porosity and permeability profiles.
Determination of the productive area is difficult to make as long as no extension wells are available. Frequently geophysical data, geological considerations and pressure measurements will help. The dip of the formation must be taken into account, in particular when it is steep. However, various graphical methods also permit the establishment of a geometrical representation of the pay horizon's rock volume and thus its calculation with a high degree of accuracy (Fig. 140).
Porosity is generally measured directly in the laboratory from cores or cuttings and then plotted in a porosity profile. These measurements are usually verified by various downhole measurements such as sonic log, neutron measurements and density log or by a determination of the formation resistance factor F.
Saturation is a major factor. The irreducible water saturation S_{wi} is best established by capillary pressure measurements carried out on cores or cuttings. In addition to this, the ARCHIE formula is also useful in this respect. An exact determination of the oil/water contact is usually difficult and requires experience.
The product of net thickness (10 m, for instance), productive area (3,000,000 m^2, for instance), porosity (25 %, for instance, or 0.25) and oil saturation (such as $1 - S_{wi} = 1 - 0.25 = 0.75$) then yields the original oil in place under reservoir conditions (in our example 5,625,000 m^3 oil). By multiplying with the formation volume factor

Table 5 Volumetric reserve calculations

Definition of productive limits of a field (A)

1. By sufficient number of appraisal wells
2. With the aid of geophysical data
   a) (geophone lowering: velocity control)
   b) ("spill point")
3. Extrapolation of pressure gradient (especially that of natural gas) and comparison with hydrostatic pressure
4. With the aid of capillary pressure measurements (determination of depth of edge water line)
5. By flowing pressure measurements over extended periods of time (reservoir limit test)
   (flowing pressure gives "echo" of field limit; practically applied only in gas fields)
6. With the aid of pressure build-up curves
   (in favourable cases, separating fault = field limit is indicated)
7. Pulse – Testing.

B_0, this volume is finally converted into "tank oil", the volume it would have on the surface. The final question then is how much of this volume is recoverable, that is, how great the recovery factor is.

The determination of recoverable reserves in natural gas-field is carried out in similar fashion. A knowledge of reservoir pressure and temperature then permits the gas volume to be converted into its value under standard conditions by considering the compressibility of the gas (z-factor). In closed reservoirs without edge water drive, the recovery factor is high and eventually depends on the required pipeline pressure. Average recovery in gas reservoirs without edge water drive is of the order of 70 to 85 % of the gas in place. In fields produced by edge water drive ultimate recovery is lower. This is due to the fact that non-homogeneous or heterogeneous reservoir rocks cause gas to be isolated by encroaching water.

5.1.1.1 Recovery Factor

The recovery factor expresses as a percentage (or as a decimal figure) how much of the original "oil/gas in place" can be economically recovered. Sometimes this amount is also stated in m^3 (bbls) recoverable per ha (acre), especially as long as the size of the field is not yet known.

The fact that it is not possible to recover all of the oil (or gas) in place has mainly two reasons:

a) Part of it can not be recovered for physical reasons. It is the oil (or gas) retained by "capillary forces" or, in other words, the oil (or gas) prevented from production because the relative permeability has dropped to zero. This is described in more detail on page 82. This volume also includes the oil (or gas) in isolated pore spaces (page 8).

b) Part of the oil (or gas) can not be recovered for economic reasons. When production costs exceed the proceeds, production becomes unprofitable and the field is shut in. It follows that fluctuations of proceeds and costs affect and modify the reserve situation.

According to a US-statistic the recovery factor in water drive fields is between 20 and 80 % (average 52 %), in gas drive fields it is between 10 and 60 % (average 33 %, Fig. 107). In the Federal Republic of Germany the average recovery factor for all oil-fields is around 29 % (without the planned thermal recovery measures).

The following values are taken from some fields in the Federal Republic of Germany:

Field 1	Lower cretaceous sandstone	water drive	44 %*
	Dogger sandstone	water drive	63 %**
	Upper oolite horizons	gas drive and weak water drive	19 %
Field 2	Dogger sandstone	expansion of oil + dissolved gas	9 %
Field 3	Dogger sandstone	gas drive-gravity	48 %
Field 4	Jurassic sandstone	gas caps, gas liberation + partial water drive	9 %
Field 5	Wealden/Jurassic horizons	gas drive	14 %

* Mobility ratio 35 : 1
** Mobility ratio 1 : 1

5.1.1.2 Classification of reserves

When reserves are estimated by the volumetric method the reliability of calculations increases with the amount of data available. In the beginning, these data are still scarce and they increase as the field is drilled up. Much the same applies when evaluations of the expectations by secondary recovery measures are made. In this case also, the degree of reliability increases with the number of observations.
Therefore, it is common practice to classify reserves according to their "reliability". Such a classification (generally dividing the reserves into "proven" and "probable") is mainly a matter of definition and agreement. The classification by *Arps* (1962)[1] was selected because it appears to be increasingly used (table 6).
Some companies try to overcome this "problem of uncertainty" by "discounting" the reserves, for instance in the following manner: proven reserves = 100 %, probable reserves = 50 % of calculated values (discounted reserves).
A very neatly defined classification is applied by the countries of the Soviet Block. As it is uniformly applied, it offers excellent possibilities of comparison (refer to *Shdanov,* 1963)[7].

5.1.1.3 Expectation curves

Everybody knows that reserve calculations are subject to uncertainties, inaccuracies and even to psychological factors. Although this is inevitable and lies in the nature of things we have to try to keep these sources of error as small as possible.
A valuable aid in this respect is the expectation curves, which use a statistical approach and, with the help of the Gaussian distribution curve, enable the reserves to be read for any probability. Although

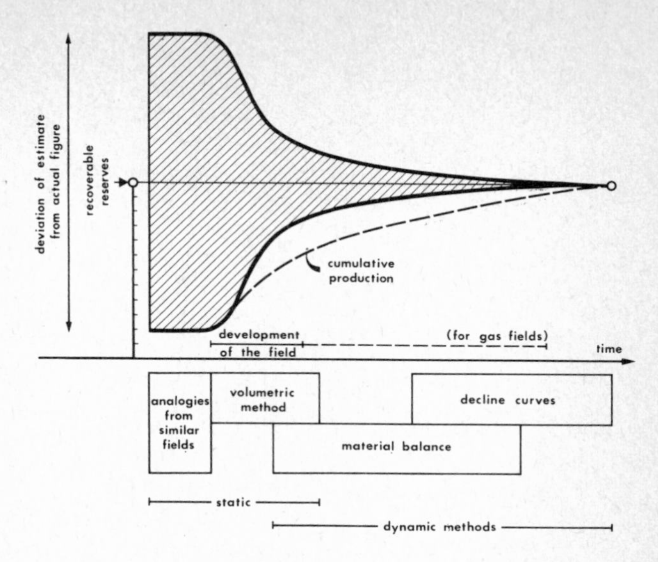

Fig. 141 The diagram shows which methods of reserve determination are applied at what times during the "life" of a field. Also shown is the deviation of estimate from the actual figure (which is known "exactly" when the field is depleted. (After *Arps*[1])

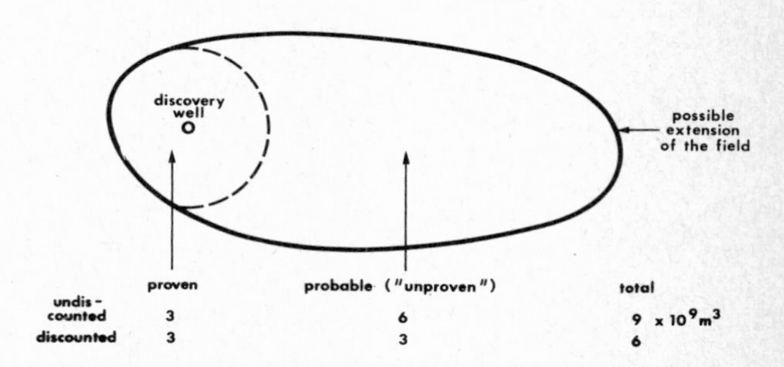

Fig. 142 In volumetric reserve calculations, uncertainties are great, especially if only one or a small number of wells has been drilled. Reserves are therefore classified as proven and probable ("unproven"). Probable reserves are often "discounted" – in the present example by 50 % –, which means that only 50 % of the volumetrically determined reserves in the "probable area" are taken into account

Table 6 Classification of Petroleum Reserves (after *Arps*[1])

Energy source	Degree of proof	Development status	Producing status
Primary Reserves Those reserves commercially recoverable at current prices and costs by conventional methods and equipment as a result of natural energy inherent in the reservoir	*Proved* Primary reserve which have been proved to a high degree of probability by production from the reservoir at a commercial rate of flow or in certain cases by successful testing in conjunction with favorable complete core analysis data or reliable quantitative interpretation of log data	*Developed* Proved reserves recoverable through existing wells	*Producing* Developed reserves to be produced by existing wells from completion interval(s) open to production
			Nonproducing Developed reserves, behind the casing or in certain cases at minor depths below the producing zones, which will be produced by existing wells
		Undeveloped Proved reserves from undeveloped spacing units in a given reservoir which are so close and so related to the developed units that there is every reasonable probability that they will produce when drilled	
	Probable Primary reserves which have not been proved by production at a commercial rate of flow, but-being based on limited evidence of commercially producible oil or gas within the geological limits of a reservoir above a known or inferred water table-are susceptible to being proved by additional drilling and testing		
	*Possible** Primary reserves which may exist but where available data		

* The term "possible" it increasingly used for reserves expected from new discoveries.

Those reserves, recoverable commercially at current prices and costs, in addition to the primary reserves as a result of supplementing by artificial means the natural energy inherent in the reservoir, sometimes accompanied by a significant change in the physical characteristics of reservoir fluids	*Proved* Secondary reserves which have been proved to a high degree of a probability by a successful pilot operation or by satisfactory performance of full-scale secondary operations in the same reservoir or in certain cases a similar nearby reservoir producing from the same formation	*Developed* Proved reserves recoverable through existing wells from a reservoir where successful secondary operations are in progress	Developed reserves to be produced by existing wells in that portion of a reservoir subjected to full-scale secondary operation
			Nonproducing Developed reserves to be produced by existing wells upon enlargement of the existing secondary operations
		Undeveloped Proved reserves which will be produced upon the installation and operation of a secondary recovery project and/or by the drilling of additional wells	
	Probable Secondary reserves which are thought to exist in a reservoir by virtue of past production performance or core, log, or reservoir data, but where the reservoir itself has not been subjected to succesful secondary operations		
	*Possible** Secondary reserves from reservoirs which appear to be suited for secondary operations but where available data will not support a higher classification		

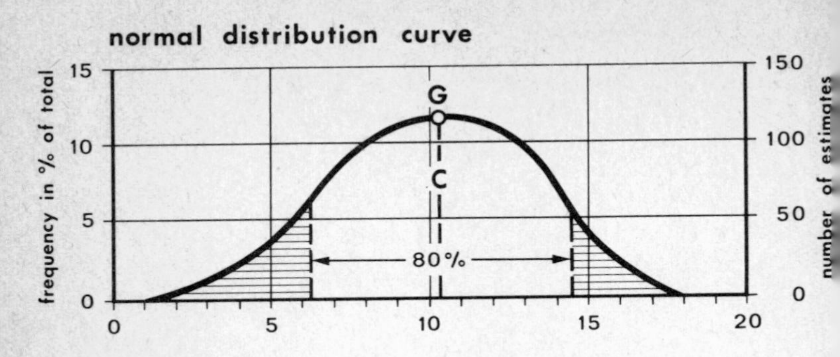

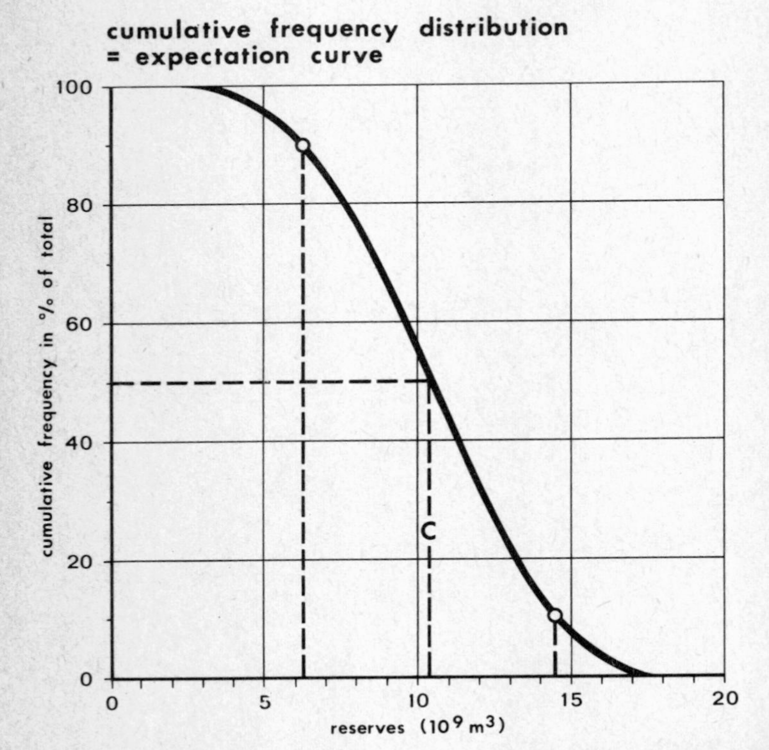

Fig. 143 When the reserves of a newly discovered field are estimated by a (very) great number of engineers, this results in a distribution curve as shown above. For a symmetrical distribution (Gaussian bell), C is the

EXPECTATIONS	EXPECTATION OF ULTIMATE RECOVERY				
	EXPECTATION OF ULTIMATE RECOVERY FROM EXISTING FIELDS				EXPECTATION OF RECOVERY FROM FUTURE DISCOVERIES
	CUMULATIVE PRODUCTION	EXPECTATION OF RESERVES FROM EXISTING FIELDS			
RESERVES	DEGREE OF PROOF	PROVEN RESERVES		UNPROVEN RESERVES	
	DEVELOPMENT STATUS	DEVELOPED	UN-DEVELOPED		
	PRODUCING STATUS	PRODUCIBLE	NON PRODUCIBLE		
	CONNECTION STATUS	CONNECTED	NON CONNECTED		

Fig. 144 Terminology of Expectation and Reserves. (After *Betz*[2])

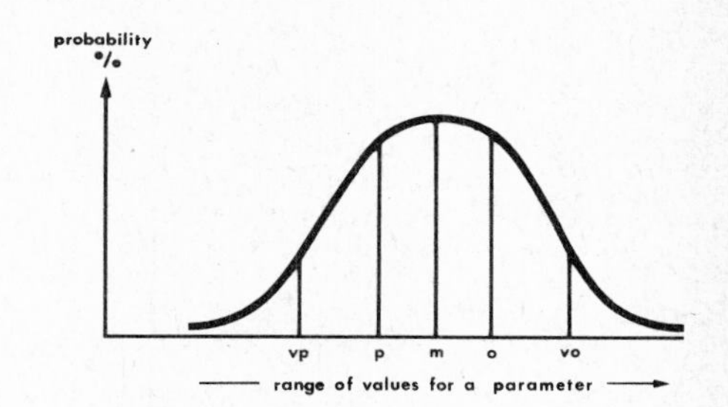

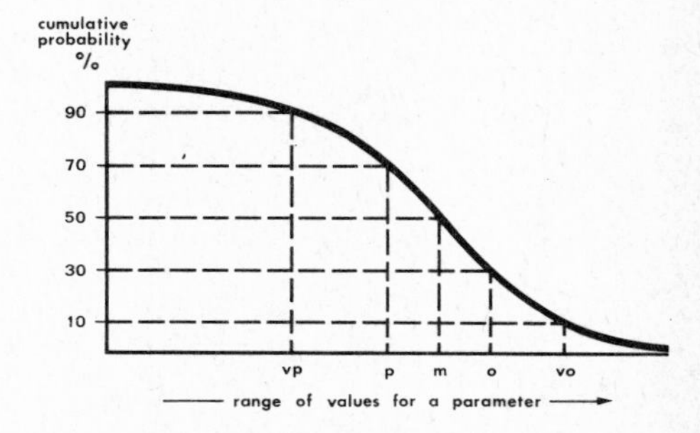

Fig. 145 Van der Laan has suggested to use only 5 ranges of values for each parameter: a "very pessimistic" one (vp), a "pessimistic" one (p), a "medium value (m), an "optimistic" (o) and a "very optimistic" one (vo). With this simplified method it is possible to use desk calculators for a quick determination. (After *v. d. Laan*[4])

---

medium value. To the left and right of it, stray areas (confidence areas) can be assumed. The above example shows that around 80 % of the estimates are within relatively narrow limits. The cumulative distribution curve is called the expectation curve of the field

A ▼ \ h ▸	6	8,5	11	13,5	16 m
20	120	170	220	270	320
29	175	246	319	391	464
35	210	297	385	472	560
41	246	348	451	553	656
50	300	425	550	675	800
km²					

Fig. 146 The example shows how two independent parameters are combined with each other and thus simplified. In the present case, they are the productive area A and the thickness h. At first A and h are multiplied. The 25 values are then reassembled into groups of 5, beginning with the smallest and ending with the greatest. From these groups of 5, the average is then formed (e.g.

$$120 + 170 + 175 + 210 + 220 = \frac{895}{5} = 179)$$

Each of these 5 averages has again a probability of 20 %. (After *Betz*[2])

this method can also be used for exploration ventures we shall restrict ourselves here to the expectations in existing fields (Fig. 144).

When a new field is discovered we have generally 4 uncertain parameters:

a) the interpretation of the seismically surveyed structure
b) the location of the edge water line
c) the character of the payrock (porosity, thickness, saturation etc.)
d) the recovery factor.

Assuming just 2 solutions for each of these 4 sources of error produces 64 different variations for the purpose of calculation the reserves, each of which has the same probability of 1.56 %. In practice, however, even more than 2 solutions have to be considered so that the number of possible variations is enormous.

Now *Van der Laan*[4] has proposed a simplified procedure which even takes into account certain subjective elements. By this procedure, probability distributions independent of each other can be computed in a simple manner. For practical application he proposed a distribution with 5 interpretations:

vo = very optimistic, a value having just a 10 % chance of being too low
o = optimistic, a value having a 30 % chance of being too low

m = the median value having equal chances of being either too high or too low
p = pessimistic, a value having a 30 % chance of being too high
vp = very pessimistic, a value having just a 10 % chance of being too high.

Each of the 5 values stated has a probability of 20 %.
When 5 figures have been established for the reserves (R_1 to R_5) each value would have, as already pointed out, a probability of 20 %. So cumulative probability for R_1 would be 90 %, for R_2 70 %, for R_3 50 %, for R_4 30 % and for R_5 10 % (Fig. 147).
When distribution is symmetrical the "reserve expectation of the field" can be read directly at the probability of 50 %. Whether the dividing line between "proven" and "unproven" is drawn at 90 % (or at 95 % or even at 85 %) is, after all, a matter of agreement. In any case the sum of proven + unproved equals the "reserves expectation of the field".

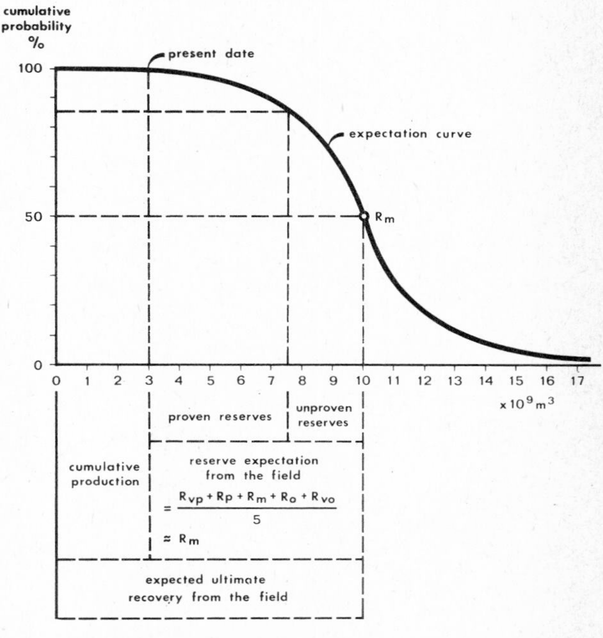

Fig. 147 For each expectation curve it is a matter of convention to define the degree of probability of the various reserve categories (classes). In the present example, a probability of 85 % was postulated to divide proven and unproven reserves. (After *Mayer-Gürr*[6])

5.1.2 "Dynamic" Methods

From a practical point of view the production decline curves should be mentioned first of all. This method is well known, and usually the rate of production (daily or monthly or yearly) is plotted on the ordinate. Sometimes it may be advisable to plot pressure, water cut or similar parameters. On the abscissa the time (in years) or the cumulative production are shown.

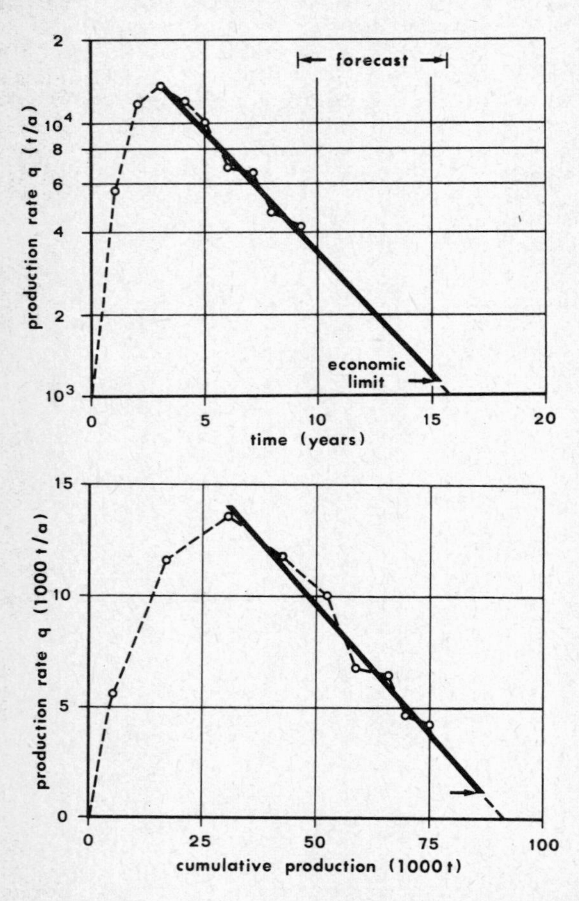

Fig. 148 For decline curves, the best forecasts are obtained when plotting the production rate q logarithmically (upper diagram) or using the cumulative production (lower diagram)

The basic assumption of such curves is of course that the performance of the field will be more or less the same in the future and, therefore, one may extrapolate the decline until the end of production. This end of production is not necessarily the physical exhaustion of the field but the economic limit of production. At this economic limit the daily production q is described by the formula $q = \frac{K}{30.4 \times P}$ where P represents the current crude oil price and K the production costs which are usually shown on a monthly basis. As both factors fluctuate it becomes evident how the volume of reserves is affected by the economic limit.
Each field has a decline curve peculiar to this field, and mathematical solutions are restricted to a few cases of special nature, namely

a) constant percentage decline
b) hyperbolic decline and
c) harmonious decline.

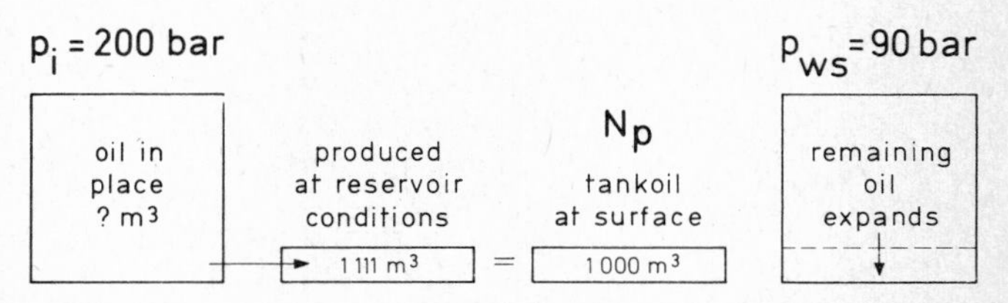

Fig. 149 An oil-field with an initial reservoir pressure (p_i) of 200 bar has produced 1000 m³ tank oil (N_p, measured under surface conditions). In the reservoir this corresponds to 1111 m³ (at a compressibility of oil (c_o) of 10×10^{-5} bar $^{-1}$). This volume withdrawn from the reservoir is replaced by the expansion of the remaining oil. From that the initial "oil in place" may be calculated (see Appendix P)

An extension and a further mathematical development of the material balance equation (refer to page 107) permits, in certain cases, a computation of N or G = the original volume of oil or gas in the reservoir. Naturally this method cannot be applied until the field has been produced for some time (some 10 % of original reserves). Generally speaking the following cases may be considered:

a) the number of unknown variables in oil-fields with a gas drive and in gas-fields (also without water drive) is limited to just one, namely N = the original oil volume or G = the original gas volume.
b) 2 unknown variables appear in the case of gas cap-fields (without water drive) namely N and m. (m = the ratio of the gas cap

volume to the oil volume). The two unknown variables in gas-fields with a water drive are G and W_e (= the invaded edge water) and in oil-fields with a water drive they are N and W_e.
c) 3 unknown variables (N, m and W_e) finally appear in oil-fields with a gas cap and an edge water drive.

The cases with just one unknown variable can be easily solved as for instance shown in a simplified calculation of gas reserves neglecting compressibility of connate water and payrock according to the formula:

$$G = \frac{G_p}{1 - (\frac{p_n}{p_i} \frac{z_o}{z_n})}$$

G = Gas initially in place
G_p = Cumulative production at reference date n
p_i = Initial reservoir pressure (bar)
p_n = Reservoir pressure at reference date n (bar)
z_o = Initial real gas factor (at pressure p_i)
z_n = Real gas factor at reference date n (at pressure p_n)

This calculation results in a minimum volume, i.e. the reserves contained in the drainage area of the well.
In the other cases 1 or 2 unknown variables must be determined by other means. Usually this is N or G, which is determined volumetrically and then entered into the material balance equation. It is more difficult to determine the water influx W_e by other means (for instance either empirically from the degree to which outpost wells have gone to water, or with the help of the *Van Everdingen-Hurst* method or the viscous fingering theory, or by the *Marsall* method) and then to enter it into the material balance equation so as to permit a computation of N. In the end, however, various material balance equations drawn up at regular intervals will yield useful values both for N and W_e. By trial and error, other factors (such as permeability irregularities) may, and must, subsequently be entered as well.
Reservoir Limit Tests may be an additional check in evaluating oil or gas reserves. When production is started in an oil- or gas-field, pressure conditions around the well are disturbed. With the production rate being constant, a "pressure funnel" develops which has a flowing pressure p_f in the borehole opposite the pay horizon and a reservoir pressure (or closed-in pressure) p_s at the boundary of the "pressure funnel"; i.e. at the border of the well's drainage area ($p_s > p_f$). This period is called the period of "non-steady-state" flow and may be divided into a) early "non-steady-state" and b) late "non-steady-state". The division becomes evident in the

behaviour of the flowing pressure, if this is plotted against time in a graph.
The pressure funnel around the well does not expand further if the influx from the outside (e.g. by encroaching edge water) equals the volume withdrawn from the well. In this case one speaks of a period of "steady-state" flow. A reserve calculation based on this method would only show the reservoir content within the drainage area.
When the drainage radius of the well reaches the reservoir limit, the period of "non-steady-state" flow changes to a period of "semi-steady-state" flow. From then on, the pressure decline per unit of time is independent of location; that is to say, only the level of the pressure funnel will change, no longer its form. In this case, the oil/gas reserves of the reservoir can be determined fairly exactly.
Each of the 4 flow periods can be recognized and analyzed in a diagram showing flowing pressure plotted against time. With the aid of special formulae, the following can be calculated: effective permeability, skin effect, reservoir pore volume, oil/gas in place and reservoir radius.
According to *Cunningham* and *Nelson*[3] a pressure build-up curve (see page 127) can be used as well for the purpose of calculating reserves in simple cases according to the formula:

$$G = \frac{0.1752\, p_{ws}\, q\, \Delta t}{zm}$$

(where Δt = time since shut-in and m = inclination of the straight-line part of the curve). Conditions: semi-steady state and drainage area = productive area of field.

5.2 Some considerations on economics

The petroleum engineer must inevitably familiarize himself with the economic implications of his work, for it is he, after all, who creates (almost exclusively) the financial means enabling his company and the producing industry to carry on exploration, discover new reserves, and maintain the whole administrative organization of an oil company. It is therefore advisable to discuss at least briefly a number of basic definitions that are essential for the petroleum engineer's planning work.

a) Investments

Investments are expenditures for long-life assets, which are financed partly from a company's capital (contributions by the owners, retained earnings and depreciations) and partly by borrowing. Examples of long-life assets in our case are all wells including their technical outfit (casings, sub-surface installations, all kinds of surface installations) and the flowlines to the central gathering station. Investments further include the entire technical outfit of a

field, i.e. all gathering, measuring and storage facilities, as well as all technical installations required to transport the oil or gas to the refinery or consumer; moreover, all auxiliary structures that are necessary to maintain operations in a field, e.g. roads, water and electricity supply, buildings, repair shops, warehouses, etc., etc.
By investments we also mean the supply of working capital to a producing unit, which permits it to procure the necessary materials for starting up and operation (stockkeeping).
These investments are normally concentrated in the early development stages of a field, especially in the case of oil-fields. As we shall see later, investments for the development of gas-fields continue to be required until the end of the plateau period.

b) Operating expenses and costs

In order to maintain operations, i.e. production in an oil- or gas-field, money must be spent on wages, salaries, materials, fuel, services, etc., etc. (operating expenses). To keep a check on the operating expenses of a company, its accounting department uses the so-called cost accounting method, which covers not only the above-mentioned operating expenses but also certain fixed costs, such as administrative overheads and depreciation on long-life assets. Under such a cost accounting system operating expenses may be either grouped into cost centers, i.e. functions, as they logically arise in production operations as follows:
subsurface costs, e.g. well cleaning, replacement of tubings and packers, paraffin removal, etc.
surface costs, i.e. all operating costs arising in connection with a well's surface installations
costs of gathering in the field
costs of oil or gas dehydration in the field
all costs of measuring and (remote) control operations
costs of natural gas desulphurization
costs of oil and gas transportation.
Operating expenses may also be divided, however, by so-called cost categories [or elements], e.g. personnel, materials, fuel, repairs, service and other contractors, etc.
One purpose of cost accounting is to determine for a given period of time the materials and services consumed by a certain product including its use of long-life assets. The calculation of genuine operating expenses is of importance for the "cash flow" analysis discussed in more detail below. It need not be specially emphasized that operating expenses arise as long as the field produces. They are partly related to the production volume and the number of wells, but are partly more or less independent of the activity in the field.

c) Depreciation

All means of production created by investments, i.e. in our case wells, pipelines, dehydration plants and all other technical installations that may be used over extended periods of time, are "written off" according to certain rules, and taxable income is reduced by the amounts so written down. The rules are fixed by the taxlaws of each individual country. For example, a well costs DM 1,000,000 including all equipment and may be written off by ten equal annual instalments; income from production say in the 5th year is DM 250,000 and operating expenses are DM 80,000. This results in the following calculation:

income	DM 250,000
less operating expenses	DM 80,000
less depreciation	DM 100,000
= taxable income	DM 70,000
less taxes*	DM 35,000
= profit	DM 35,000

* assumed tax burden 50 % of taxable income

The purpose of this scheme is to ensure that new capital is formed to replace the depreciated assets, thus securing a company's continued existence.

Since the investment expenditure for the well arose in the first year, "genuine" costs in the present year are only operating expenses and taxes; thus:

income	DM 250,000
less operating expenses	DM 80,000
less taxes	DM 35,000
= surplus	DM 135,000

For the purposes of internal cost calculation, one may depart from the legally established rules and, for instance, write an asset down over its economic lifetime. If this in 15 years for the abovementioned well, depreciation can be distributed over this period (i.e. DM 66,000/year).

d) Taxes

An important expense item (whose importance is, unfortunately, constantly increasing) are taxes payable to the government. A simplified income tax calculation has already been shown. Taxes levied by the government include also royalties. As they consume large amounts of money, taxes are of paramount importance for the economics of every project. They occur under a variety of names (income tax, production tax, turnover tax, royalty special taxes, etc.) and are related either to profit, to turnover or to production. They are different in every country, and the petroleum engineer should carefully familiarize himself with the lax laws of the country which affect his planning.

e) Cash flow

The first step in the economic evaluation of a project is to compare cash flows, i.e. cash-out and cash-in. If cash-in during a given period of time (say one year) is greater than cash-out, one speaks of a cash surplus. If the opposite is the case, one would have a cash deficit.

f) Present day value

If one pays DM 1000 into a savings bank account at the beginning of a year, the credit balance at the end of the year will be DM 1100, assuming an interest rate of 10 %. Now if I can expect to receive the sum of DM 1100 after one year, this amount, if available today, would be worth to me "only" DM 1000.

As is shown by this simple example, present-day value calculation is nothing but an inversion of interest calculation. In profitability analysis, it is used to show the present-day value of an income that will only be realized at a later date. The higher the interest rate and the later the income realization, the smaller the present-day value. It is easily conceivable, how important this consideration will become say for an oil-field producing over 30 years.

To evaluate the economics of a project, a number of yardsticks have been developed using different approaches making it possible to compare the economics of different projects. The most important of these are:

a) present-day value of cash surpluses, which is calculated according to the following formula:

$$K_{pv} = \sum_{i=1}^{n} K_i \cdot \frac{1}{(1 + r)} i$$

where:

K_{pv}	= present-day value
K_i	= cash surplus or deficit in i-th year
$i=1, \ldots, n$	year
r	= interest rate

The present value of cash surpluses (or deficits) may now be compared with that of alternative projects. The project with the highest cash surplus is the most profitable one.

b) Earning power
Earning power or DCF rate is the interest rate r = x, for which the discounted cash surpluses and deficits are zero. If a project is supposed to yield a profit, x must be positive. The earning power of alternative programmes is compared; the project with the highest earning power is the most profitable one.

c) Payout time
The number of time intervals (mostly months or years) is determined for which the sum of cash surpluses and cash deficits is zero. This is the time within which the project is amortized.

5.3 The development of oil-fields

The following considerations are based on the assumption that an oil reservoir is developed and produced by *one* operator (or at least by "unit operations") and that there is no restriction from the outside. In other words, the basic task is how to develop and produce an oil-field so that the reservoir as a whole gives the maximum final recovery and the optimum economic result. This is different from the philosophy of the early days of the industry when all thinking was focussed on the individual well.
If we consider first of all the economic point of view, the problem is best characterized by taking the two extreme ways of possible development:

a) if we develop and produce an oil-field by drilling only a few favourably located wells (from the geological and structural point of view), the investment costs will be extremely low.
Furthermore the operating costs per unit of time will also be low, and production will be at a low level, but it will continue for a long time and, therefore, the present value of the cash-in will be small.

b) The other extreme would be to develop the field by many (closely spaced) wells and produce it at high rates. In this case the capital expenditure for wells and equipment will be extremely high. The operating costs will also be very high (manpower, energy, dehydration, repairs and maintenance); the same is true for the cash-in flow and the field will be exhausted in a relatively short time. This results in a high present value of cash-in.
It will be readily seen that it is a matter of pure calculation to find a point where the two extreme considerations converge in an optimum number of wells to be drilled. This optimum development plan will result in maximum economic success. This is shown in Fig. 150 which shows a curve derived from a schematic simplified calculation.
In addition to the economic considerations there are a lot of other factors which play their part in the future development planning

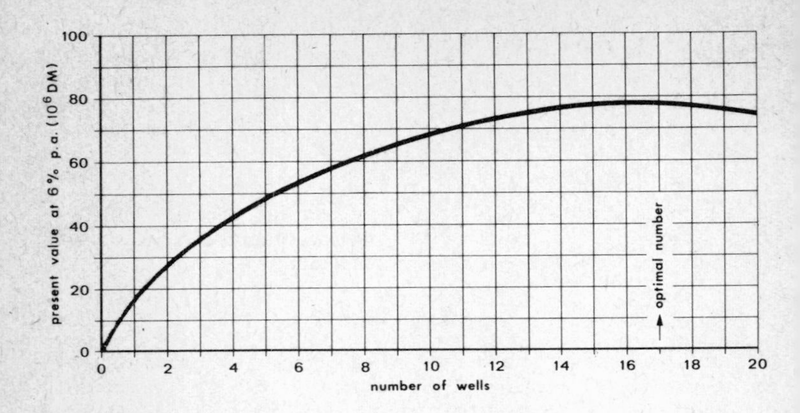

Fig. 150 The number of wells which leads to a maximum economic success can easily be calculated. In this example 17 wells to be drilled will result in an optimal cash surplus

of an oil-field. There are, among others, mainly the geological, the structural and the reservoir point of view which may be summarized as "reservoir geometry". By way of example the following items may be mentioned: the type of structure controls the drilling-up to a great extent. A normal well-shaped anticline calls for a different development plan than a salt dome flank type of field or a heavily faulted monocline. Sometimes these blocks are separated by tight faults so that each block must be regarded and handled as a single reservoir. Another geological factor governing the development plan is the frequent heterogeneity of the reservoir rock especially regarding its permeability. And finally the type of drive mechanism will have a decisive influence on the drilling programme: expanding gas caps will lead to a different drilling pattern than for instance encroaching edge water. Anyhow, it will always take a certain number of extension wells to get sufficient information on the geological, structural and petrophysical features of the newly discovered field and on the performance of the reservoir.

As far as the physical side of the problem is concerned, it is well known that prior to the intensive scientific work on hydrocarbon reservoirs the attention of producers had been more or less focussed on the single well. Combined with legal practices which contributed to cutting a reservoir into small pieces this led to close well spacing and to the production of each well at its full capacity. By this competitive approach waste of oil became inevitable and led to a number of superfluous wells. A misinterpretation of "Cutlers Rule" formulated in 1924 ("doubling the distance between wells doubles the ultimate production per well and halves the ultimate recovery per acre") led to the conclusion that the number of wells affects the efficiency with which the reservoir energy is used and that the ultimate recovery of an oil-field depends on the spacing of the wells drilled. Today we know that the amount of recoverable oil depends on the type of structure, the type of reservoir performance, the properties of the rocks and fluids and the development and operating plans and does not depend on the density of spacing of wells. As described in the previous chapters fluids flow to the well as long as there exists a pressure potential in the reservoir, and this flow occurs over long distances.

5.4 Development of gas-fields

The principles governing the development of a gas-field are entirely different from those applicable to oil-field development. As we

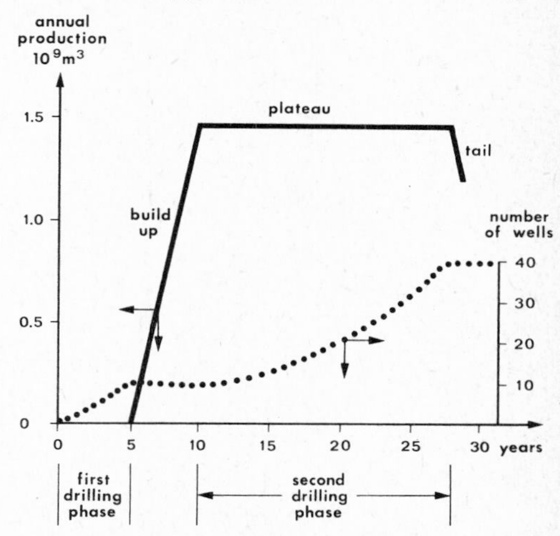

Fig. 151 Schematic representation of the development and production of a gas-field. Normally, after the discovery, a number of wells will be drilled to outline the field and to collect the information necessary as a basis for future planning. These wells of the "first drilling campaign" will suffice to produce the gas required for the so called "build-up period". In order to maintain the "plateau production" it will take additional capacity wells in order to compensate for the decreasing capacity of the existing wells. This will lead, towards the end of the plateau period, to an ever increasing number of wells until, with the end of the "second drilling campaign", the "tail production" sets in with decreasing production rates reflecting the declining reservoir pressure

shall see, this fundamental difference also essentially affects the reservoir engineer's tasks as well as the investment planning. The development of a natural gas-field depends to a large extent on matching the pattern of the sales contract, i.e. the customer's requirements with the performance of the field and the art of petroleum engineering.

If natural gas, say from a newly discovered reservoir, is offered for sale, it is usually the first negotiating aim to spread the available volumes over a certain period of time which is generally a compromise between the requirements of the customer and the reservoir. The time schedule agreed upon will start with a build-up period (for pipeline construction, conversion of customer's appliances, etc.). Experience has shown that this build-up period usually lasts 2 to 5 years. Depending on the type of consumer, some 5–10 % of the field's recoverable reserves are withdrawn during the build-up period.

Normally the customer will be interested in obtaining a constant annual volume of gas over as long a period as possible. This volume is called the "plateau production". The period concerned will be

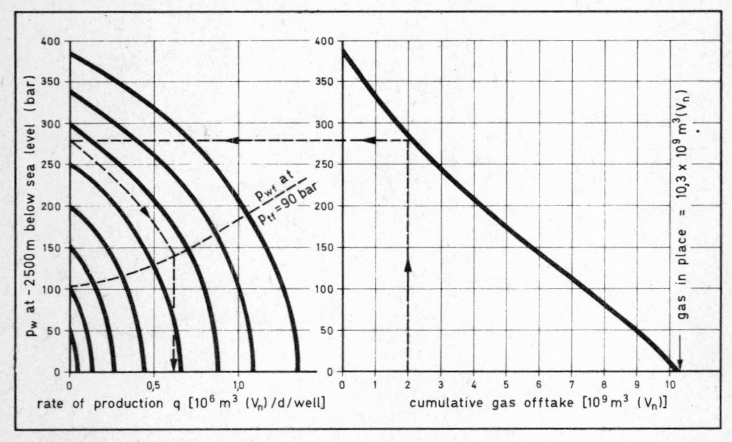

Fig. 152 With the help of such a nomogram the capacity of a well can be directly read. On the right side the pressure decline as a function of the cumulative production is shown. In the above example the reservoir pressure has dropped from its original value of 385 bar to 280 bar after a production of 2×10^9 m³. On the left side of the nomogram the daily production rate q can be read for each particular value of reservoir pressure. In our example the capacity of a well decreased from 1.1×10^6 m³ per day at the beginning to 0.6×10^6 m³ per day after a cumulative production of 2×10^9 m³ (and with a minimum wellhead flowing pressure of 90 bar for a given tubing diameter). (After *Leicht*[5])

either short with a large volume, or long with a small volume. At the end of the plateau period, some 60–70 % of the recoverable reserves will have been produced. The rest of the available natural gas volumes will be sold as the so-called "tail production".
The sales profile agreed upon during the negotiations is the essential basis for all plans concerning the drilling up and technical development of a gas-field, because the reservoir pressure declines in the course of production. (For simplicity's sake, it will be assumed here that we have to deal with an "ideal gas" and that the reservoir has no water drive. The illustrations are based on this assumption). In reality, due to its various components, natural gas is not an "ideal gas" and therefore does not obey the law $PV = RT$, but rather $PV = zRT$, where z is the supercompressibility factor (see page 56).
Along with the decline of reservoir pressure, the well capacity decreases according to the formula $q = C\,(p^2_{ws} - p_{wf}^2)^n$. For practical purposes, diagrams are available which permit a direct reading of the production capacity as a function of reservoir pressure and cumulative gas production (see Fig. 152).
In general, a number of wells will be available in the first drilling phase, which is aimed at evaluating the size of the field before the start of production. These wells will suffice to meet the requirements during the build-up period and possibly the first years of the plateau period. Depending on the rate at which the reservoir pressure and, therefore, the capacities of the individual wells decline, it will be necessary in a second drilling phase to complete and put into production additional wells. The number of production wells which is ultimately required is essentially a function of the sales policy adopted. It is easily seen that this sales policy has a decisive influence on the investments required for the development of the field. (Continued on page 174)

Bibliography

[1] *Arps, J. J.:* In: *Frick, Th. C.,* Petroleum Production Handbook. McGraw-Hill Book Co., 1962.

[2] *Betz, D.:* Reservenerwartungskurve als Informationsmittel und Bewertungsgrundlage. Erdöl-Erdgas-Zschr. 86, Juni 1970.

[3] *Cunningham* and *Nelson:* I. Petr. Techn., July 1967.

[4] *Laan, G. v. d.:* Physical properties of the reservoir and volume of gas initially in place. Verh. v. d. Kon. Ned. Geol. Mijnbouk Gen. Deel 25, 1968.

[5] *Leicht, H.:* Erdöl und Kohle, May 1967.

[6] *Mayer-Gürr, A.:* Erdöl und Kohle, März 1969.

[7] *Shdanow, M. A.:* Berechnung der Vorräte von in Erdöl gelöstem Gas. Zschr. Angew. Geologie, Berlin 1960.

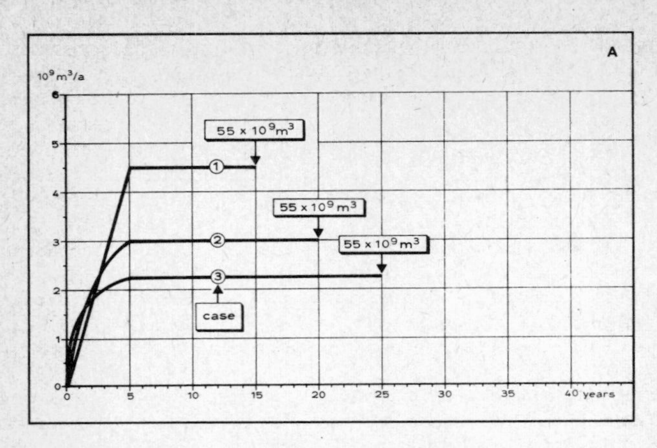

Fig. 153 When starting from the basic assumption that our negotiations with a customer will finally result in a build-up and a plateau period we can easily see that the plateau production may be either high and short (as in case 1 of Fig. A) or low and long (as for instance in case 3). In all cases 55×10^9 m³ (or about 65–70 % of the recoverable reserves) will be produced and sold

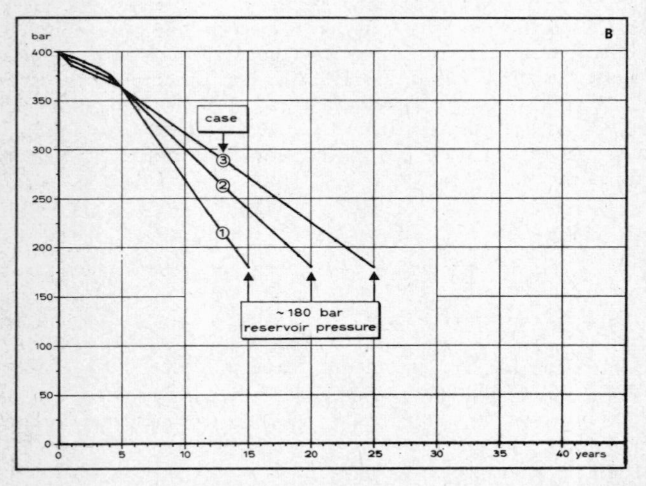

Let us further assume that the initial reservoir pressure of our field amounts to 400 bar and that (with an initial gas in place of 100×10^9 m³ and recoverable reserves of 75×10^9 m³) the pressure will drop to 180 bar after the production of the above mentioned 55×10^9 m³ (Fig. B)

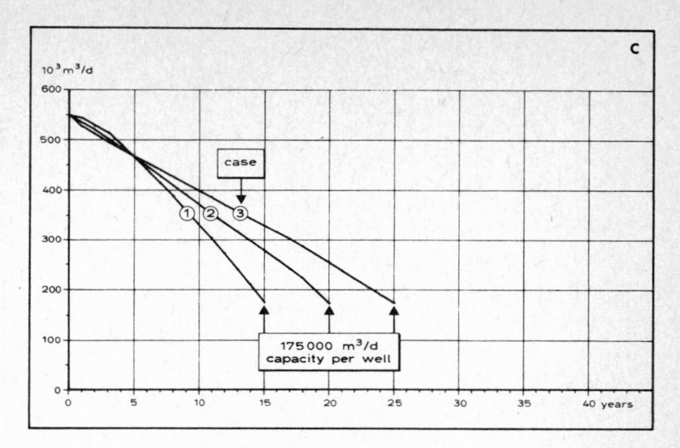

With the help of a nomogram (similar to that shown in Fig. 151) one can easily see that the capacity of a well has dropped from 550×10^3 m³/d in the beginning to 175,000 m³/d at the end of the plateau period (Fig. C)

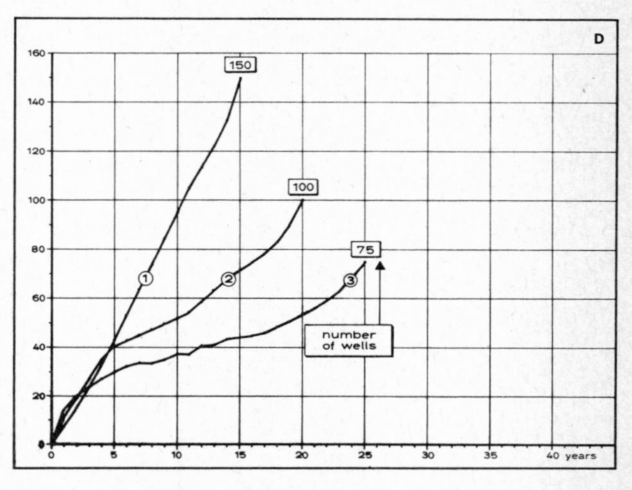

In order to fulfil our supply commitments until the end of the plateau period (and let us assume a load factor for peak delivery of 50 %) and with the above mentioned figures in mind, one can calculate that in case 1 we need 150 wells in order to be able to produce this high rate at the end of the plateau period. In case 3 with a lower production plateau we need only 75 wells (Fig. D)

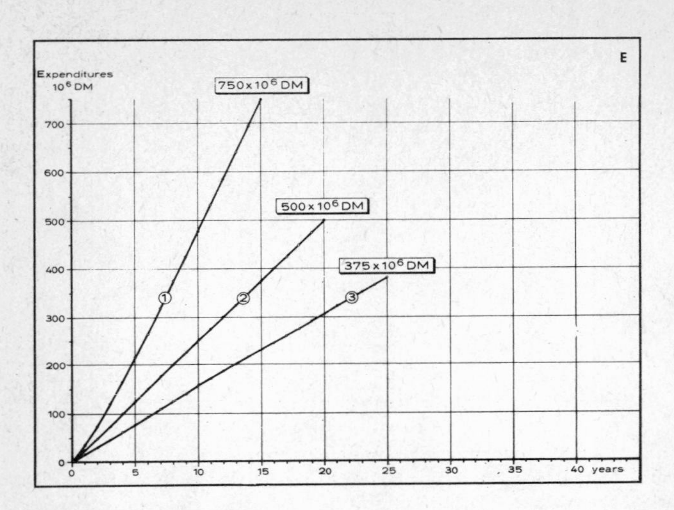

And finally: if a well costs 5 million DM including equipment it is evident to what degree the expenditures for investments are governed by the offtake pattern (Fig. E)

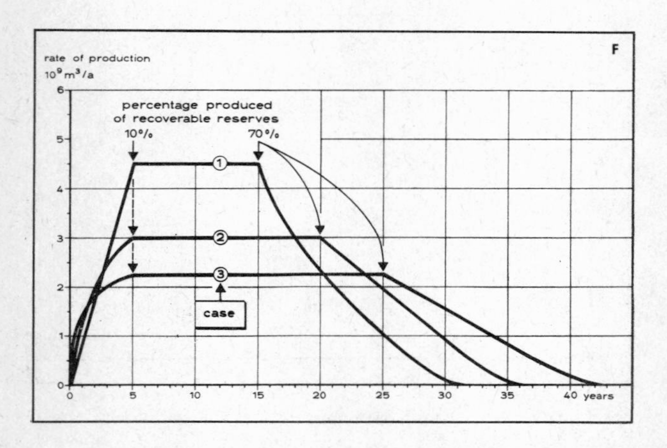

These schematic examples show to what degree the development planning for a gas-field and the costs are affected by the type of sales contract. It is, therefore, quite indispensable to consult the petroleum engineer when negotiating the sales contracts

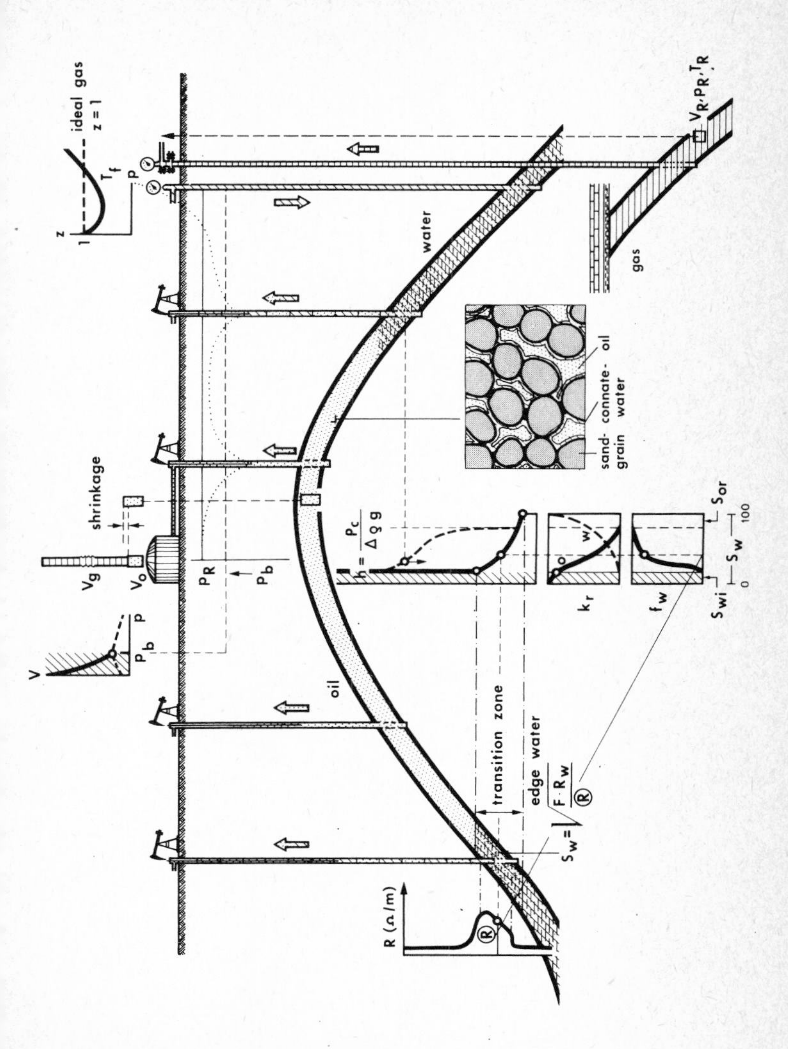

Fig. 154 (By courtesy of Dr. *G. Miessner*)

Appendix A

How to determine average porosity

Usually the porosity is measured in the lab on cores, which were taken from the reservoir rock. Fig. 155 shows e.g. 82 measurements at equal intervals in a series of about 10 m. The porosity varies between 1 and 25 %. Which average porosity should be used e.g. for reserve calculations

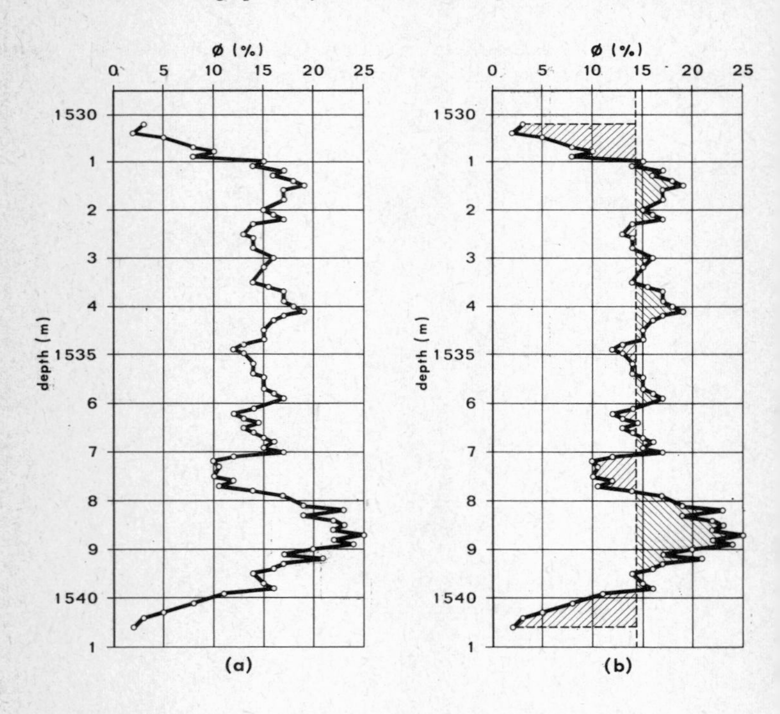

Fig. 155 In a series of about 10 m thickness, 82 porosity measurements were made at equal intervals. With the help of a "Gaussian Distribution Curve", an average porosity of 14.5 % was determined. The dashed areas to the left of this value are as large as the areas to the right of the 14.5 % line (Fig. b)

and other investigations? With the help of a distribution table and a graphical representation, this average value is easy to determine as follows:

The porosity is split up into intervals, in this case into porosity values up to 2, 4, 6 % etc. (column 1).
In column 2 there is the number of samples, 82 in total, split up according to the above-mentioned intervals.
Column 3 shows again the same distribution, this time as a percentage of samples investigated; i.e. 82 samples = 100 %.
Column 4 finally indicates the accumulated percentage frequencies.

Table 7 Distribution table

(1) intervals of porosity values	(2) number of samples	(3) % of total samples	(4) accumulated percentage frequency
2	2	2.44	2.44
4	2	2.44	4.88
6	2	2.44	7.32
8	3	3.66	10.98
10	3	3.66	14.64
12	7	8.53	23.17
14	17	20.73	43.90
16	19	23.17	67.07
18	14	17.07	84.14
20	5	6.10	90.24
22	4	4.88	95.12
24	3	3.66	98.78
26	1	1.22	100.00
total	82	100.00	

Graphically there appears a "Gaussian Distribution Curve", and the maximum is between 14 and 16 % porosity. Under the 50 % value on the accumulated percentage line, one finds the average porosity to be used for practical purposes.

If this average porosity is entered in the porosity diagram, the area to the left of the 14.5 % line must be as large as the area to the right of it. (Fig. 155 b)

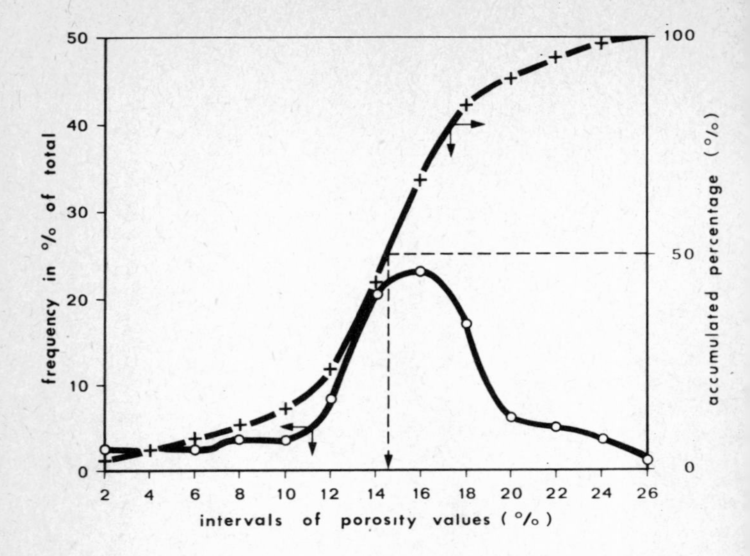

Fig. 156 The intervals of porosity values of the distribution table are plotted against "frequency in % of total" (= Gaussian Distribution Curve) and against "accumulated percentage". The 50 % value on the latter curve corresponds to the average porosity to be used for practical purposes

Appendix B

Methods for Determining Porosity*		gives:
1	on rock samples	
1.1	by liquid displacement	
1.1.1	in volumetric flask	
1.1.1.1	non-wetting liquid (Russel-volumeter)	B
1.1.1.2	wetting liquid	M
1.1.2	in Hg-porosimeter	B
1.2	by weighing with pycnometer	M
1.3	by buoyancy	
1.3.1	with non-wetting liquid	B
1.3.2	with wetting liquid	M
1.4	by saturation of the pore space	
1.4.1	with wetting liquid	P
1.4.2	with non-wetting liquid	P
1.5	by compression and expansion of gas	
	(compression volumeter	M
	expansion volumeter	B
	Washburn-Bunting porosimeter)	
1.6	on thin sections	
1.6.1	geometrically	P
1.6.2	statistically	P
2	by well logging	
2.1	from F and m	P
2.2	Neutron log	P
2.3	Sonic Log	P
2.4	Density Log	P
2.5	Micro Log	P
2.6	Microlatero Log	P

B = Bulk Volume; M = Matrix; P = Pore space

* See also: *Amyx, J. W., D. M. Bass jr., R. L. Whiting:* Petroleum Reservoir Engineering. Physical Properties, New York 1960

Appendix C

Determination of porosity on thin sections

Since, for naturally occurring porous materials, the porous structure is of a spatially random nature, the plane porosity of a random section must be the same as the volumetric porosity.

There are two methods to measure porosity: a geometrical and a statistical one. Both methods use enlarged photographs of thin sections. The pores are specially marked by impregnating the rock coloured resin.

The geometrical determination is made by planimetering.

The total area porosity is determined by the statistical method in the following manner: a pin is dropped in a random manner on an enlarged photograph of a thin section and a hit in the pore space is recorded as a positive result P. The quotient n_p/N, in which n_p means the number of positive random pin tosses of N tosses made, approaches the porosity value ϕ for large N. It shows the probability for a hit in the pore space.

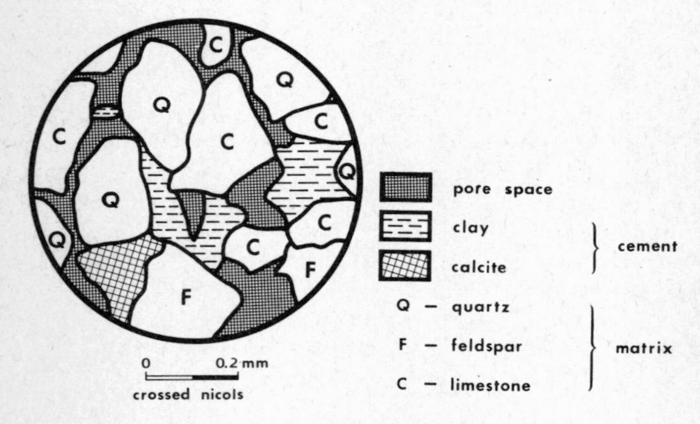

Fig. 157 Thin section of a reservoir rock

Fig. 158 Comparison of porosity values of an Upper Triassic sandstone of a German gas-field

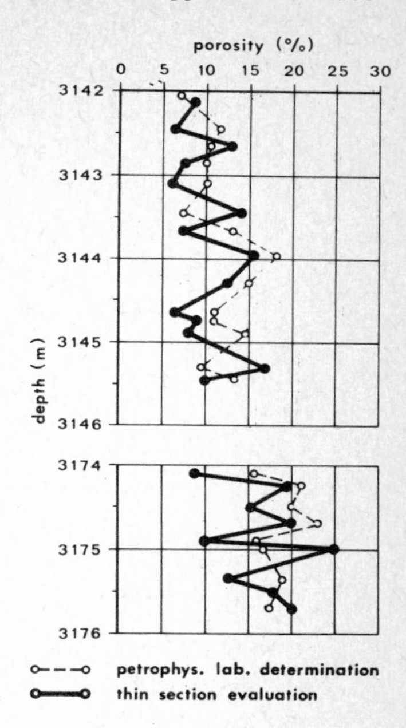

Appendix D

X-ray diffraction analysis

In crystalline solids, the structure consists of crystal building blocks in the form of atoms, molecules or ions. These elements join together according to space patterns to form a crystal lattice structure. In such a lattice system, the equidistant lattice sheets are arranged on top of one another. These sheets consist of the above-mentioned equally spaced structural elements (lattice building blocks or lattice points), and when laid together form a lattice sheet network.

W. H. and *W. L. Bragg* (father and son team who in 1915 received the Nobel Prize for physics) discovered, in 1913, that a ray passed through or reflected by a lattice network will give off interference or diffraction pattern in the form of reflections. X-rays are only reflected by the crystal lattice, however, at angles meeting the equation

$$2\,d \sin \Theta = n\lambda.$$

In the X-ray diffraction analysis method, the wavelength λ of a monochromatic X-ray diffraction diagram. Using Bragg's equation one can then figure out the value of d. And it is this d-value that identifies the crystalline substance in question.

Appendix E

Notes on Capillarity

Capillary rise of water in a tube of capillary dimension is demonstrated and the amount of P_c in an oil/water column is presented.

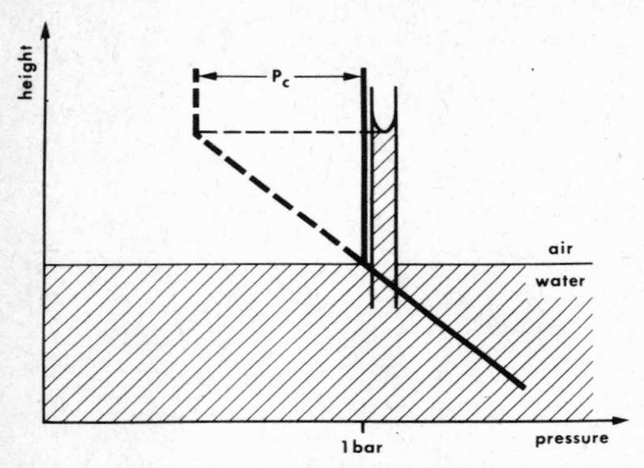

Fig. 159 The pressure outside the capillary tube is indicated by the thick solid line (when neglecting the weight of air). The pressure inside the capillary tube is in accordance with the dashed line. Pressure diferential $= P_c$

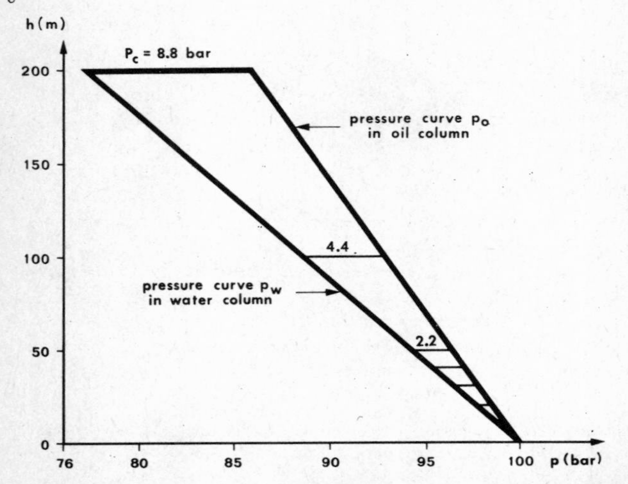

Fig. 160 shows the same for water/oil. In this example the capillary pressure P_c is 8.8 bar at a height of 200 m above the "free waterlevel"

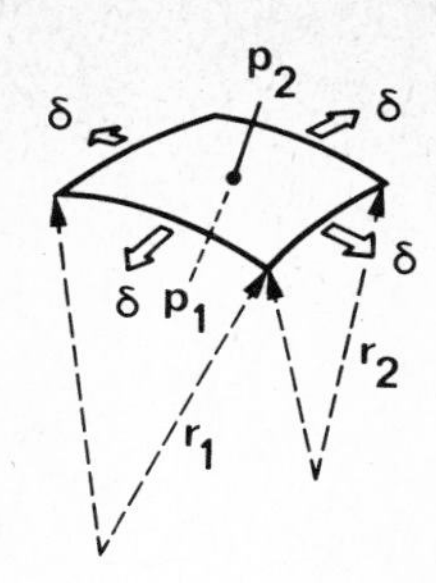

normal pressure at curved interface

$$p_1 - p_2 = \Delta p = \delta \left(\frac{1}{r_1} + \frac{1}{r_2}\right)$$

r_1, r_2 main radii of curvature

δ interfacial tension

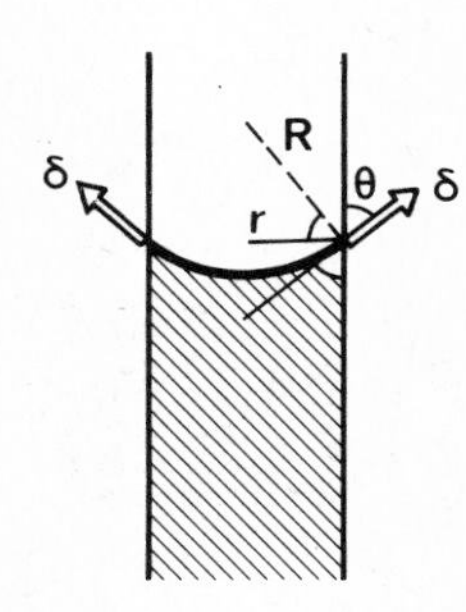

to capillaries, the following applies

r radius of capillary

$r_1 = r_2 = R$ main radius of curvature

$$R = \frac{r}{\cos\theta}$$

$$\boxed{\Delta p = \frac{2\delta\cos\theta}{r} = P_c}$$

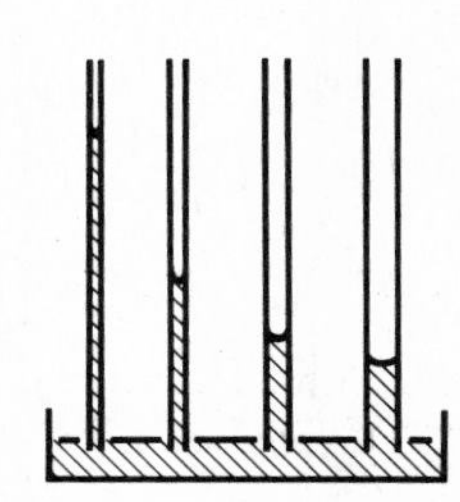

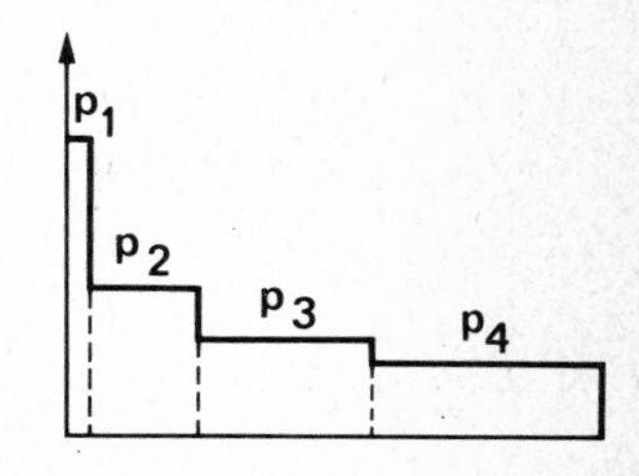

Fig. 161 Capillary parameters (By Courtesy of Dr. *G. Miessner*)

Appendix F

Hg-injection method to determine P_c

Fig. 162 shows the process in 3 steps, for which the following relationships apply:

$$P_1 = P_{c1} = \frac{2\,\sigma \cos \Theta}{r_1}$$

$$P_2 = P_{c2} = \frac{2\,\sigma \cos \Theta}{r_2}$$

$$P_3 = P_{c3} = \frac{2\,\sigma \cos \Theta}{r_3}$$

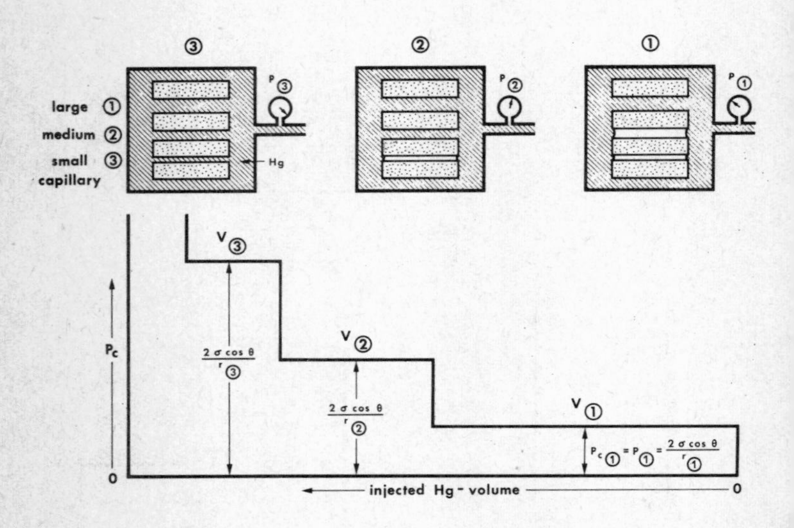

Fig. 162 Injection of Hg into a rock species with large, medium and small capillaries. Compare also Figure 43 in which the rock sample is shown before Hg invaded the large capillary. Dashed = Hg; dots = sand grains; white = pores ("capillaries")

The values which during this lab measurement are obtained for an air/mercury/rock system must now be converted to a gas/oil/rock system by using the following conversion factors:

Table 8

system	Θ	$\cos \Theta$	σ dyn/cm	$\sigma \cos \Theta$
air/water/solid	0°	1	72	72
air/mercury/solid	140°	–,766	480	– 370
oil/water/solid	0°	1	35	35

$$\frac{P_c \text{ air/water/solid}}{P_c \text{ air/mercury/solid}} = \frac{72}{-370} = -\frac{1}{5{,}1}$$

$$\frac{P_c \text{ oil/water/solid}}{P_c \text{ air/mercury/solid}} = \frac{72}{-370} = \frac{1}{10{,}5}$$

Appendix G

Natural Gas Formation Volume Factor B_g

The formation volume factor expresses the relationship of the volume under reservoir conditions to the volume under standard conditions. In the case of natural gas, it is more practical to work with the reciprocal value; by multiplying by this reciprocal, one receives from a specified reservoir volume the volume in m³ (V_n).
The reciprocal of the formation volume factor is determined by using this simple equation.

$$\frac{1}{B_g} = \frac{p_W \cdot 273.16}{z \cdot T_r \cdot 1.033}$$

or simplified

$$\frac{1}{B_g} = \frac{p_W \cdot 264.434}{z \cdot T_r}$$

Symbols:

B_g – Gas-formation volume factor (without dimensions)
p_W – Reservoir pressure (bar)
z – Super compressibility factor at pressure p_W
T_r – Reservoir temperatur in °K = °C + 273.16

Appendix H

Klinkenberg effect

The phenomenon that for gases under *low* pressures the volume flux per unit area is greater is described by Klinkenberg by the following equation:

$$k_L = k_\infty \left(1 + \frac{b}{p_M}\right)$$

The constant b depends on the average capillary radius r and the free path length λ of the gas molecules, i.e. $b \approx \lambda/r$.

$$p_M = \frac{p_{\text{before sample}} - p_{\text{behind sample}}}{2} \cdot k_\infty$$

corresponds as extreme value $\lim p_M \rightarrow \infty$ to the permeability for a liquid of inert behaviour (= permeability as rockspecific constant).

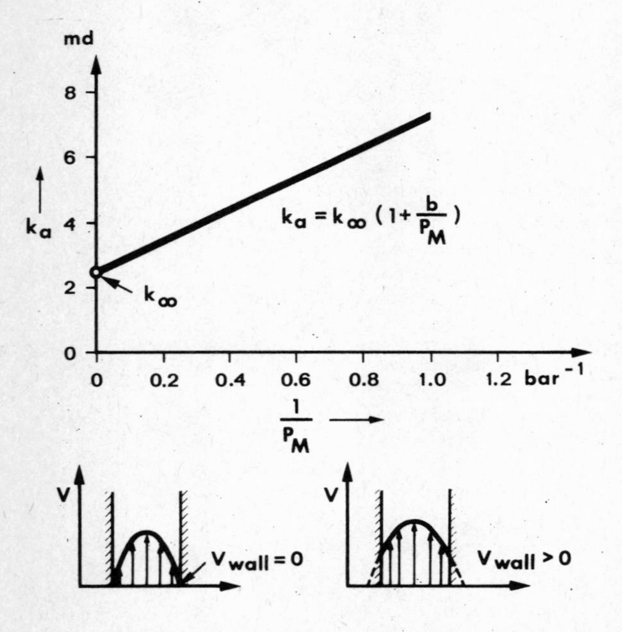

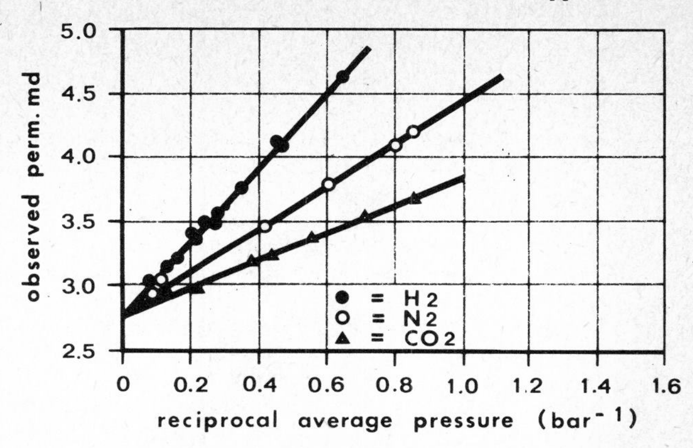

Fig. 164 Shows the Klinkenberg effect to be more pronounced for H_2 than for N_2 or CO_2

Fig. 163 Shows a "Klinkenberg straight line" obtained experimentally and illustrates the increasing deviation of k_a (permeability for air) with decreasing pressures. For liquids or gases under high pressure, where the free path length λ of the gas molecules is very small compared to the pore diameters, the boundary condition $V = o$ is met at the pore walls (V = velocity of flow), as the molecules adhere to them. With decreasing gas pressure, the free path length λ of the gas molecules becomes greater and greater and finally approaches the smallest capillary diameters. As gas pressure decreases, therefore, on the velocity of flow (volume flux per unit area) there is increasingly superimposed a velocity component resulting from molecular movement, which acts in the direction of pressure decline. The average free path length of gas molecules in the direction of pressure decline (less collisions due to smaller concentration of molecules) is greater than in the direction of pressure increases. This effect may be described with the boundary condition $V_{wall} > 0$ as slipping along the pore walls or as an enlargement of the effective capillary radius

Appendix I

Swelling phenomena

If the matrix of a porous medium contains "swellable" substances (e.g. sandstones with clay), it will be observed that the permeability depends on the electrolyte content of an aqueous (or on the dipole moment of a non-aqueous) pore filling. A detailed discussion of intracrystalline swelling processes, as they occur e.g. in montmorillonite, and of interfacial phenomena – hydratation, dipole adsorption – would exceed the scope of this survey.
The sensitivity of the rocks containing swellable clay substances regarding their water permeability depends very much on the texture and distribution of clays. A rock in which the sand grains are covered by fibrous clay overgrowths is less sensitive than a rock containing a network of porefilling clays.

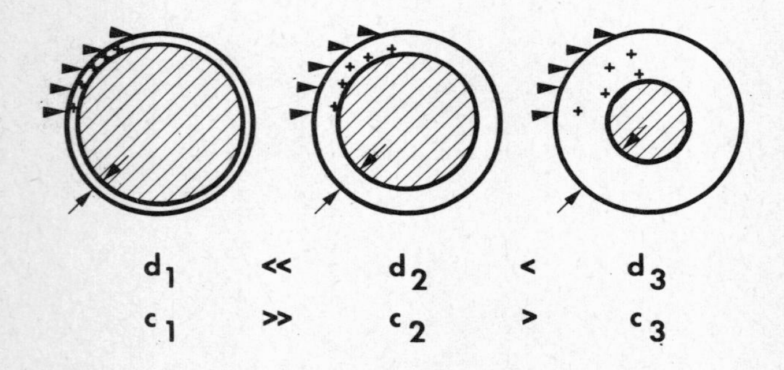

Fig. 165 Clay Swelling. The reduction of water permeability with decreasing electrolyte content is demonstrated.
The "effective cross section" of a pore depends on the electrolyte concentration. For a high concentration C_1, the thickness d_1 of the "immobile" interface is small, the volume of intracrystalline swellable substances about normal, and the water permeability usually not far below the "Klinkenberg permeability" K_∞. With decreasing electrolyte content C_2 and C_3, the interface thickness and the effective volume of intercrystalline swellable substances increase; in this way, the effective pore radius becomes smaller and smaller. This means (according to Hagen-Poiseuille's Law) a substantial permeability reduction especially where electrolyte contents are small

Table 9 shows the result of a lab measurement on a sandstone sample, which contains about 10 wt.-% montmorillonite.

Table 9

k_{air} = 12 md, ϕ = 17.2 %

NaCl-conc. (wt.-%)	k_{water} (md)
26	9.7
10	7.8
3	5.6
1	2.1
0.5	0.085
0.2	0.008

Appendix K

Frontal-advance-rate formula

By solving the frontal-advance-rate formula of Buckley / Leverett by step for each saturation and each f-value, the saturation in a linear flooding experiment is obtained. Let us illustrate the problem by an example.

When a piston is moved forward in a pipe of a certain cross section, a volume of water situated in front of the piston will migrate to the right at the same speed and over the same path. In Fig. 166 example A shows this process as well as the relevant saturation diagram when the front of the flood has covered a path Δx.

water volume $\Delta Q = 1 \cdot A$ (in which A = the pipe's cross section and Δt = time)

Velocity $$V = \frac{1}{\Delta t} = \frac{\Delta Q}{\Delta t A}$$

Now if the same piston moves the same volume of water not through a pipe, but through a rock of a certain porosity, the path the volume of water covers in the porous rock is obviously longer than that in the pipe. Its length depends on the rock's porosity. This is shown in example B.

Injected water volume $\Delta Q = \Delta x \cdot A \cdot \phi$ (ϕ = Porosity)

$$\Delta x = \frac{\Delta Q}{\phi A}$$

$$V_{(front)} = \frac{\Delta x}{\Delta t} = \frac{\Delta Q}{\Delta t \phi A}$$

It will be readily seen that, in the porous rock, flowing speed exceeds piston injection speed by the factor $\frac{1}{\phi}$.

When the rock's pore space is not "empty" but saturated with oil, flow processes are complicated still more. The flood front will then no longer advance like a piston. Water saturation will decrease as a function of relative permeability and of viscosities in the direction of flow. f_w, relative water flow in a given cross-section of the porous rock, is subject to similar factors. This is shown in example C. It will be readily seen that the "points" of low water saturation S_w or the "points" of small f value have migrated "farthest" (or, in other words, "quickest").

Each cross section ("slice") of the porous rock shows a certain saturation S_w (x, t) and the relevant f value f_w (S_w(x, t)). The f value describes relative water flow through this "slice". Injection speed V, however, also changes as compared with example B. Therefore $V_{Sw} = \frac{Q}{t \phi F} G_{(Sw)}$. The factor G describes how much faster these slices of equal water saturation move as compared with the velocity of example B.

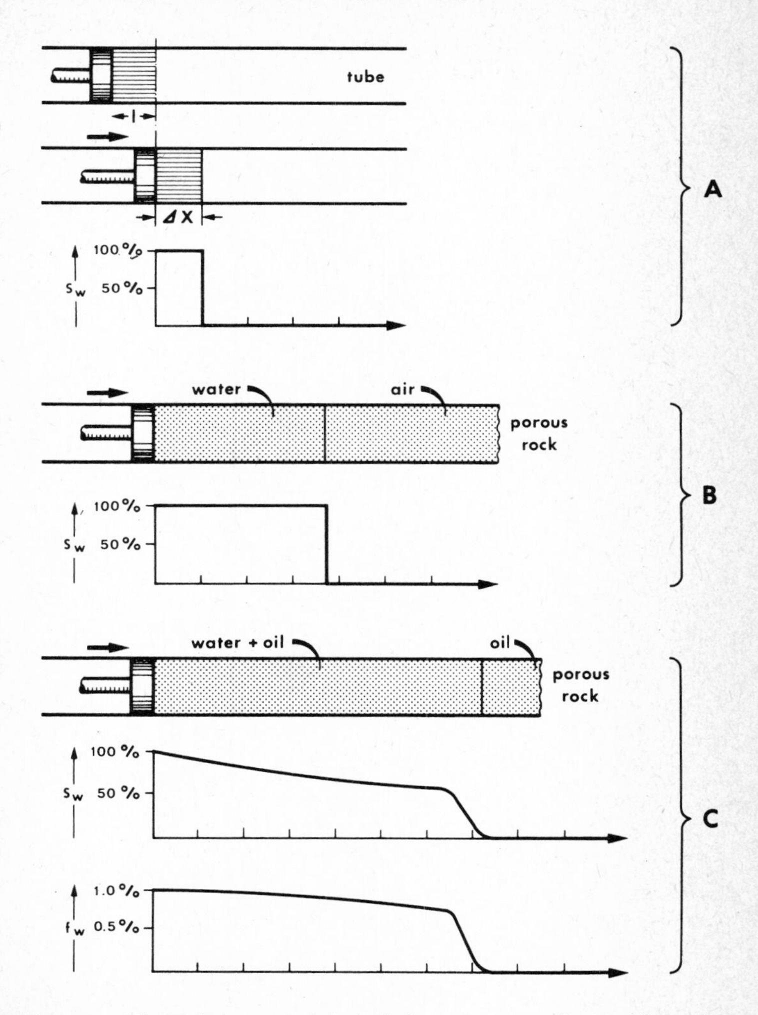

Fig. 166 Schematic representation of the movement of water through a pipe, a dry porous, and an oil-saturated rock

Appendix L

Example of a simple material balance and oil-in-place calculation for an undersaturated oil-field which still produces above the bubble point

It is assumed that the field in question has been producing for a certain period of time, after which a material balance calculation is made. The following quantities must then be known from measurement in an autoclave or in the field:

original reservoir pressure	p_i	$= 200$ bar
present reservoir pressure	p_{ws}	$= 90$ bar
bubble point pressure	p_b	$= 80$ bar
formation volume factor	B_o	$= 1.111$
oil compressibility	c_o	$= 10 \times 10^{-5}$ bar^{-1}
production to date, tank oil	N_p	$= 1000$ m^3

Pressure still being above the bubble point, the following has occurred so far in the reservoir:

As a result of production, reservoir pressure has decreased by ($p_i - p_{ws} =$) 110 bar. The volume withdrawn from the reservoir – measured at the surface – is ($N_p =$) 1000 m^3. Within the reservoir, however, this volume had occupied a somewhat greater space, as the oil has "shrunk" on its way up to the surface (p. 48); the volume withdrawn from the reservoir is therefore $N_p \times B_o = 1111$ m^3. Due to the decline in reservoir pressure by ($p_i - p_{ws} =$) 110 bar, the original oil-in-place N has expanded, since undersaturated crude oil at pressures above the bubble point expands by ($c_o =$) 10×10^{-1} per 1 bar of pressure decline. Thus, the oil remaining in the reservoir now occupies a volume of ($N \times 10 \times 10^{-5} \times 110$). The volume withdrawn by production is therefore replaced by expansion of the oil that has remained in the reservoir. In other words: the expansion of the remaining oil is equal to the volume produced at reservoir conditions, i.e.:

$$N \times 10 \times 10^{-5} \times 110 = 1111$$

or, as N in this equation is the only unknown quantity,

$$N = 101{,}000 \text{ m}^3.$$

The original oil-in-place of this reservoir is therefore 101,000 m^3. In order not to complicate the example unnecessarily, expansion also of the connate water, compression of the reservoir rock and corresponding reduction of the pore space were not taken into account.

Appendix M

Example of an isochronal test in a gas well

Table 10

Flowing time (h)	Shut-in time (h)	p_{ws} (bar)	p_{wf} (bar)	q (m^3/h)	$p_{ws}^2 - p_{wf}^2$	point in Fig. 122
–	72	412.0	–	–	169 744 = p_{ws}^2	–
1	–	–	388	1200	19 200	1
2	–	–	381	1200	24 583	2
3	–	–	376	1200	28 368	3
20	–	–	340	1200	54 144	4
–	20	412.0	–	–	169 744	–
1	–	–	362	2400	38 700	5
2	–	–	345	2400	50 719	6
3	–	–	333	2400	58 855	7
–	3	411.7	–	–	169 497	–
1	–	–	320	4000	67 097	8
2	–	–	287	4000	87 128	9
3	–	–	266	4000	98 741	10

Figure 122 shows the graph of points 1 to 10 on a logarithmic scale and the construction of the family of parallels. Figure 167 illustrates the estimate of the asymptotically stabilized C value.

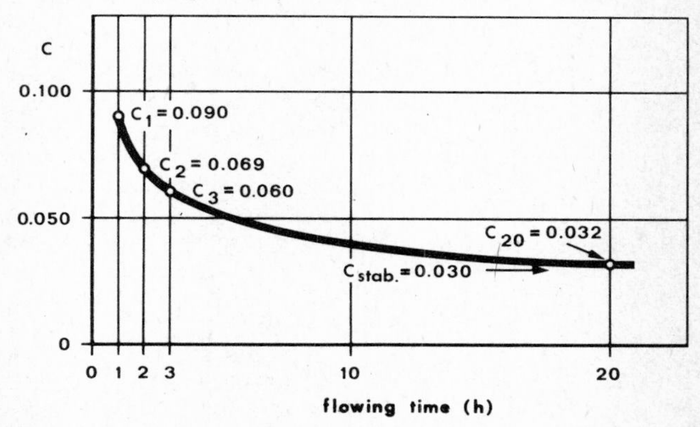

Fig. 167 Determination of the stabilized capacity coefficient C from a isochronal test. The way to find this value is to construct straight capacity lines from the flowing pressure measured after certain periods (see Fig. 122). The C values of such families of lines are then plotted against flow time. As shown in the above figure they asymptotically approach a value we may consider to be the stabilizied C value

Appendix N

Interpretation of a pressure build-up measurement in an oil well

The following parameters of the well are known:

$q = 4\ m^3/d$	production rate
$N_p = 775\ m^3$	cumulative oil produced
174	days since start of production
$r_w = 10$ cm	well radius
$\phi = 0.20$	porosity
$h = 25$ m	thickness
$B_o = 1.045$	oil formation volume factor
$\mu_o = 13.9$ cp	viscosity of oil
$c_o = 7.5 \times 10^{-5}\ bar^{-1}$	compressibility of oil
301 m.	depth of pressure gauge

The pseudo-time t_p = equivalent time the well was on production prior to shut-in can be calculated:

$$t_p = \frac{Np}{q} = 193.75 \text{ days} = 4650 \text{ h.}$$

Table 11

1	2	3	4	5	6	7
date	time	p_{ws} (bar)	p_{wf} (bar)	Δt_{ws} (h)	$t_p + \Delta t_{ws}$ (h)	$\frac{t_p + \Delta t_{ws}}{t_{ws}}$
April 5	17.30		5.6	0	4650	∞
„ 5	17.35	15.8		0.08	4650.08	58126
„ 5	18.00	16.1		0.5	4650.5	9301
„ 5	20.30	18.2		3.0	46.53	1551
„ 6	5.30	22.8		12.0	4662	389
„ 6	17.30	25.2		24.0	4674	195
„ 8	8.30	27.2		36.0	4713	75
„ 9	8.30	27.5		87.0	4737	54.4
„ 10	8.30	27.8		111.0	4761	42.9
„ 11	8.30	28.0		135.0	4785	35.4
„ 12	8.30	28.2		159.0	4809	30.2
„ 14	8.30	28.5		207.0	4857	23.5
„ 16	8.30	28.7		255.0	4905	19.2
„ 18	8.30	28.9		303.0	4953	16.3
„ 21	8.30	29.1		375.0	5025	13.4

Δt_{ws} = time after well is shut in (h)
p_{ws} = bottom hole pressure, static (bar)
p_{ws} = bottom hole pressure, flowing (bar)
On April 5 at 17.30, the well was shut in.

The figures within the thick line of the table 11 were calculated. The figures of column 1, 2, 3 and 4 have been measured.
The pressure build-up curve is shown in Fig. 123.
The formula for the pressure build-up is:

$$p_w = p_i - \frac{21.91 \times q \times \mu}{B_o \times k \times h} \log_{10} \frac{tp + \Delta t_{ws}}{\Delta t_{ws}}$$

where p_i = initial pressure

The values are shown within the thick line of the above table, and the curve is represented in Fig. 125.
Extrapolation of the straight-line part of the curve results in p^x = 32.0 bar.

tang α = m = bar/log cycle is
(m from 29.4 to 32.0 bar)
m = 2.6 bar/log cycle.

Permeability k can now be calculated:

$$k = \frac{21.9 \times \mu \times q \times B_o}{k \times m}$$

= 19.58 md.

Skin factor S is

$$S = 1.151 \left(\frac{p^x - p_{wf}}{m}\right) - \log \frac{7.826 \times t_p \times k}{\phi \times \mu \times c_o \times r_w^2}$$

= + 3.02

This skin factor causes an additional draw-down of $\Delta p_{skin} = 0.868 \times m \times S$ = 6.82 bar.

The productivity index PI is

$$PI = \frac{q}{p^x - p_{wf}}$$

= 0.152 m³/d/bar.

The capacity of the well is
= PI × p_s = 4.86 m³/d.

The theoretical PI of the well is (without skin factor)

$$PI_{theor} = \frac{q}{p^x - p_{wf} - \Delta p_{skin}}$$

= 0.204 m³/d/bar.

The theoretical capacity is

= $PI_{theor} \times p_s$
= 6.5 m³/d.

The "damage ratio" DR is

$$DR = \frac{PI_{theor}}{PI}$$

= 1.34.

Appendix O

Flowing pressure at different production rates

Russell's formulae are (in terms adapted in the F. R. of Germany):

$$k = -\frac{21{,}92\, q_1\, \mu\, B}{b\, h} \tag{1}$$

$$S = 3{,}71 + \frac{0{,}00524\, kh\, [p_{wf}(\Delta t) - p_{wf}(t)]}{(q_1 - q_2)\, \mu\, B} +$$

$$+ 1{,}15 \left(\frac{q_1}{q_1 - q_2}\right) \left[\log\left(\frac{t + \Delta t}{\Delta t}\right) - \log\left(\frac{kt}{\Phi\, c\, \mu\, r_w^2}\right) - 4{,}12 + \right.$$

$$\left. + \frac{q_2}{q_1} \log\left(\frac{13.230\, k\, \Delta t}{\Phi\, c\, \mu\, r_w^2}\right)\right] \tag{2}$$

$$p^* = p_{wf}(t) + \frac{21{,}92\, q_1\, \mu\, B}{kh} \left[\log\left(\frac{kt}{\Phi\, c\, \mu\, r_w^2}\right) + 0{,}89 + 0{,}87\, S)\right]$$

Let us illustrate Russell's method by an example from the German oil-field Steimbke-Ost. Figure 168 shows that the well produced $q_1 = 0.86\ m^3/h$ through a 4 mm bean, while through 2 mm bean the production rate decreased to $q_2 = 0.31\ m^3/h$. At q_1 the flowing pressure was (at a depth

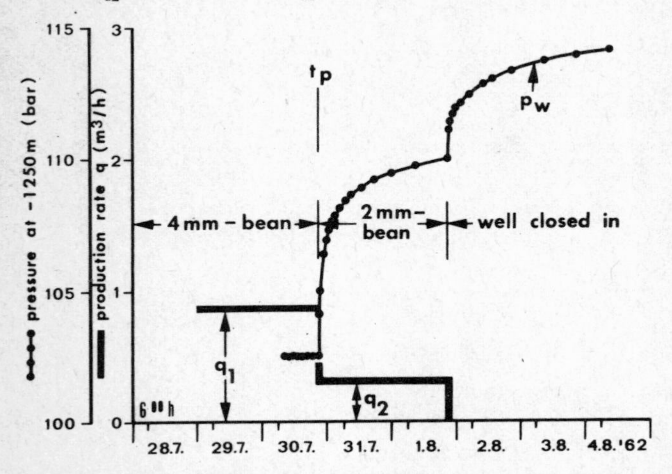

Fig. 168 Flowing pressure at different production rates in a well of the Steimbke oil-field, Germany. The production rate was decreased from q_1 to q_2, and finally the well was closed in. The bottom hole pressures were continuously recorded

of 1250 m) 102.4 bar; when the production rate had decreased to q_2, the flowing pressure p_{wf} rose, rapidly at first, then more slowly. Shortly before the well was closed in, it reached a value of 109.8 bar. The well was then closed in and a normal pressure build-up was recorded.

Parameters of the well were:

q_1	= original production rate	= 241.6 cm^3/sec.
q_2	= production rate after reduction	= 85.6 cm^3/sec.
c_o	= compressibility of oil	= 10.3 $\times$ 10^{-5} bar^{-1}
ϕ	= porosity	= 0.15
μ	= viscosity	= 2.4 cp
h	= thickness	= 500 cm
r_w	= radius of well	= 8.4 cm
B_o	= oil formation volume factor	= 1.108
t_p	= time since last shut-in	= 30.348 $\times$ 10^6 sec.

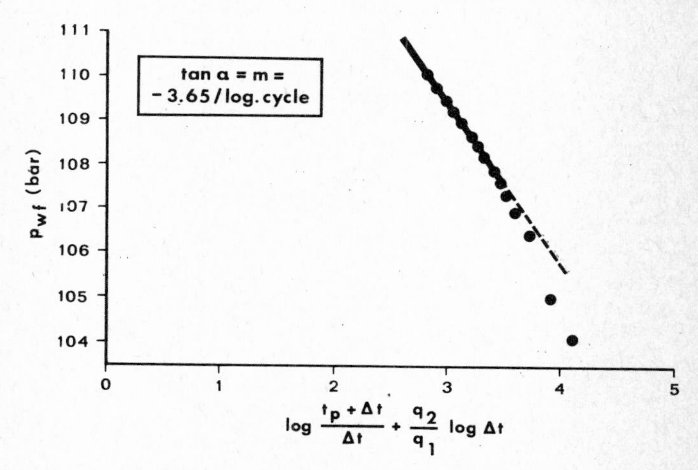

Fig. 169 Interpretation of the flowing pressures of Fig. 168. From this drawing, the static pressure p^x, the skin factor S and the permeability k can be determined

The flowing pressure p_{wf} is plotted against log

$$\frac{t_p + \Delta t}{\Delta t} + \frac{q_2}{q_1} \log \Delta t.$$

This is shown in Figure 169. (Δt being the time since reduction of the production rate, in seconds.) Part of the curve is a straight line, the extrapolation of which to log

$$\frac{t_p + \Delta t}{\Delta t} + \frac{q_2}{q_1} \log \Delta t = 0$$

yields the static pressure p*. Tang α = m is 3.65 bar/log cycle.
A comparison with the results from the pressure build-up showed satisfactory conformity:

	from flowing pressure	from pressure build-up
permeability k, md	665	666
skin factor S	– 4.28	– 4.40
p*, bar	122.3	121.3.

Appendix P

Determination of the original "oil in place" by material balance in an undersaturated oil reservoir without water drive

The basic equation is:

$$\underset{(1)}{N_p B_o} = NB_{oi} \left[\underset{(2)}{(c_o (p_i - p_{ws})} + \underset{(3)}{\frac{(c_f + S_w c_w)(p_i - p_{ws})}{(1 - S_w)}} \right]$$

(1) = volume produced
(2) = expansion of oil left in the reservoir
(3) = expansion of connate water and compression of the pore volume.

By a mere rearrangement of the original equation, one obtains the formula for a determination of the oil in place N:

$$N = \frac{N_p B_o}{B_{oi} \left[c_o (p_i - p_{ws}) + \frac{(c_f + S_w c_w)(p_i - p_{ws})}{(1 - S_w)} \right]}$$

N_p	= cumulative oil produced	= 1,000 m³
B_o	= oil formation volume factor at present pressure	= 1.111
B_{oi}	= oil formation volume factor at initial pressure	= 1.099
c_o	= compressibility of oil	$= 10 \times 10^{-5}$ bar⁻¹
p_i	= initial reservoir pressure	= 200 bar
p_{ws}	= present reservoir pressure	= 90 bar
c_f	= formation (rock) compressibility	$= 4 \times 10^{-5}$ bar⁻¹
c_w	= water compressibility	$= 5 \times 10^{-5}$ bar⁻¹
S_w	= water saturation	= 0.25
p_b	= bubble point pressure	= 80 bar

Note: Both N and N_p are expressed as tank oil volume.

The original "oil in place" is:

$$= \frac{1{,}000 \times 1.111}{1.099 \left[10 \times 10^{-5} (200-90) + \dfrac{(4 \times 10^{-5} + 0.25 \times 5 \times 10^{-5})(200-90)}{(1-0.25)} \right.}$$

$$= \frac{1{,}111}{1.099 \left(1{,}100 \times 10^{-5} + \dfrac{5.25 \times 10^{-5} \times 110}{0.75}\right)} =$$

$$= \frac{1{,}111}{1.099\,(0.011 + 0.0077)} = 53{,}932 \text{ m}^3 \approx 54{,}000 \text{ m}^3$$

Without considering (3), the result would be:

$$N = \frac{1{,}111}{1.099 \times 0.011} = 91{,}901 \text{ m}^3 \approx 92{,}000 \text{ m}^3$$

From this comparison, a weakness of the method is clearly seen: in most cases, the available data are not sufficiently accurate to permit an exact determination of the oil in place. This applies above all to the compressibility values and the formation volume factors, but in particular to the pore space compressibility, whose determination is much more difficult than that of the compressibilities for oil, gas and water. Greater accuracy is usually only obtained after repeated material balance calculations at various intervals during the producing life of a reservoir: the accuracy increases with the stages of production. This is true for both oil and gas reservoirs.
Therefore, at an early stage of production, a volumetric determination of the reservoir content will be preferable.

References

The following list gives a survey of the comprehensive books on "Petroleum Engineering" which contain further references, mostly published before 1970. Recent references on individual aspects of the subject are quoted at appropriate passages in the text.

[1] *Amyx, J. W.; Bass Ir. D. M.; Whiting, R. L.:* Petroleum Reservoir Engineering. Physical Properties. McGraw-Hill Book, Cy. Ltd., New York – Toronto – London 1960.

[2] *Bentz, A.; Martini, H. J.:* Lehrbuch der angewandten Geologie. Zweiter Band, Teil 1, 3.2. Erschließung und Ausbeutung von Erdöl- und Erdgasfeldern (*A. Mayer-Gürr*). Ferd. Enke-Verlag, Stuttgart 1968.

[3] *Burick, E. J.:* Properties of Petroleum Reservoir Fluids. John Wiley & Sons Inc., New York 1957.

[4] *Calhoun, Ir. J. C.:* Fundamentals of Reservoir Engineering. The Oil & Gas Journal & University of Oklahoma Press, 1960.

[5] *Collins, R. E.:* Flow of Fluids through Porous Materials. Reinhold Publishing Corp., New York 1961.

[6] *Craft, B. C.; Hawkins, M. F.:* Applied Petroleum Reservoir Engineering. Prentice-Hall Inc., Englewoods Cliffs, N. J. 1959.

[7] *von Engelhardt, W.:* Die Bildung von Sedimenten und Sedimentgesteinen. E. Schweizerbart, Stuttgart 1973.

[8] *Frick, Th. C.* (Editor in Chief): Petroleum Production Handbook. Vol. II Reservoir Engineering. McGraw-Hill Book Cy. Ltd., New York – Toronto – London 1962.

[9] *Katz, D. L.:* Handbook of Natural Gas Engineering. McGraw-Hill Book Cy. Inc., New York – Toronto – London 1959.

[10] *Matthews, C. S.; Russell, D. G.:* Pressure Build up and Flow Tests in Wells. American Inst. of Mining, Metallurgical and Petroleum Engineers, Dallas 1967.

[11] *Pirson, S. J.:* Oil Reservoir Engineering. McGraw-Hill Book Cy. Inc., New York – Toronto – London 1958.

[12] *Szilas, A. P.:* Production and Transport of Oil and Gas. Akadémiai Kiadó, Budapest 1975.

Register

A

Absolute Permeability 81, 83, 87
After-Production 130
Antisotropic 78
Aquifer 68, 109
Archie's Formula 24, 30
Associated Reservoir Volume 112
Atomic Subsurface Blows 27
Autoclave 47
Average Reservoir Pressure 110

B

Back Pressure Test 125
Ball's Pressure Sentry 120
BET-Method 24
Bottom Water 72, 116
Bourdon Tube 119 ff.
Bubble Point 47
Bubble Point Line 52
Bubble Point Pressure 47, 50, 53, 68, 73, 107
Build-Up Period (Gas Sales) 171 ff., 196
Bulk Volume 10 ff.

C

Capacity of Wells 121 ff., 147, 172
Capillarity 3, 9, 75, 102, 184 ff.
Capillary Pressure 34 ff.
Capillary Pressure Curve 4, 39
Cap Rock 44
Carbonate Rocks 5, 8
Cash Flow 166, 168 ff.
Casing Scheme 60
Classification of Reserves 154 ff.
Clay Swelling 190
Closed Reservoir 68
Closure 91
Coalification 8
Compaction 6, 13
Compressibility 10, 50, 72
Condensate Field 50, 53
Connate Water 40, 84 ff., 102, 108
Continuous Flowmeter 144
Cost Categories 166
Cost Centers 166
Critical Envelope Curve 51
Critical Point 51, 52
Critical Production Rate 99, 100
Cutler's Rule 170

D

Damage Ratio 197
Darcy 19, 78 ff., 112 ff., 125
DCF Rate 169
Decline Curves 148, 162 ff.
Deficit 167 ff.
Densimeter 145
Depreciation 165, 167 ff.
Detour Factor 24
Development (of Fields) 147, 169 ff., 171 ff.
Development Plan 1
Dew Point Line 52, 53
Diagenesis 6, 8, 10
Discounted Reserves 154
Displacement Pressure 38, 41, 45
Displacement Theories 93 ff.
Displacing Medium 93
Dissolved Gas Drive 68 ff., 103
Drainage Area 79, 112, 133, 135
Drainage Volume 112
Draw Down 121, 125
Drilling Plan 147
Dry Gas Reservoir 50, 54
Dynamic Method (Reserves) 162

E

Earning Power 169
Earnings 165 ff.
Economics 165 ff.
Edge Water 72, 74, 93 ff., 170
Edge Water Line 41, 68, 96
Effective Permeability 81, 127
Effective Pore Compressibility 18, 75
Electrical Models 112 ff.
Electrical Resistivity 27 ff.
End Effect 86
Engineering 1
Estimation of Reserves 147 ff.
Expectation Curves 154 ff.
Expenditures 165 ff.
Exploration 1
Extended Material Balance 109

F
f-Functions 86 ff., 104
Flowing Pressure 121 ff., 198 ff.
Flow of Fluids 77
Flow Tests 140 ff.
Flush Production 71
Formation Resistivity Factor 24, 28 ff.
Formation Volume Factor 49, 108, 129, 187
Fossil Pressures 65
Free Water Level 37, 41, 92
Front (Frontal Advance) 96, 97, 98, 192 ff.
Funicular Saturation 40, 84

G
Gas Cap 70 ff., 74, 93, 170
Gas Cap Drive 70 ff.
Gas/Oil Ratio 69 ff., 72, 73, 134
Gas-Water Contact 41, 142, 149
Gaussian Distribution Curve 154 ff., 179 ff.
Geopressures 61
Geothermic Depth Gradient 66 ff.
Grain Pressure 18, 58
Grain Sorting 9, 10, 23
Gravity 74 ff., 100

H
Heterogeneous Flow 77
Hewlett-Packard Pressure Recorder 120
Hg Injection Method 37, 102, 186
High Pressure Zone 63, 64
High Resolution Thermometer 144
Homogeneous Flow 77
Hügel Pressure Recorder 120, 122
Humble Formula 30
Hydropressures 61
Hydrostatic Pressure 59
Hysteresis Effect 33, 86

I
Ideal Gas 53, 54, 56, 173
Imbibition 33
Inhomogeneous 41
Injectivity Index 124
Inner Surface Area 23
Insular Saturation 84
Interfacial Tension 52
Interference Test 137 ff.
Investments 165 ff., 169
Irreducible Water Saturation 40, 82, 84, 89, 102
Isochronal Test 126, 195
Isothermal Pressure Descent 51
Isotropic 78

J
Jamin Effect 44

K
Klinkenberg Effect 21, 80, 188 ff.
Knudsen Flow 21, 80
Kozeny-Carman Equation 24, 27

L
Laminar Flow 21, 79
Load Factor 175

M
Maihak Pressure Recorder 120
Marsal Method 95, 99, 164
Material Balance 56, 82, 107 ff., 148, 163, 194
Mathematical Reservoir Simulation 112 ff.
Matrix Compressibility 17
Matrix Volume 10
Measurement of Porosity 11, 181 ff.
Mechanical Engineering 1
Migration 44
Mobility 86 ff.
Mobility Factor 86 ff.
Mobility Ratio 87 ff.
Montmorillonite 80
Multiple Phase Flow 81

N
Non-Steady State Flow 164

O
Oil (Gas) in Place 107 ff., 112, 124, 135, 142, 148 ff., 194, 200
Oil-Water Contact 41, 59, 97, 143
One Phase Flow 79
Open Flow Test 124
Open Reservoir 73
Operation Costs (Expenses) 166 ff., 169

Operations 1
Original Gas/Oil Ratio 50
Overburden Pressure 58
Overhead 166 ff.

P
Pack Flowmeter 145
Parallel Flow 81
Payout Time 169
Peak Delivery (Gas Sales) 175
Pendular Saturation 40, 84
Permeability 4, 19 ff., 41, 129, 134, 140, 170, 197
Perm System 20
Petroleum Engineering 1
Petrostatic Pressure 16, 58 ff.
Phase Diagram 52, 54
Piezometric Level 60
Plateau (P_c) 38, 42
Plateau Period (Gas Sales) 166, 171 ff.
Pore Compressibility 10, 14 ff., 108
Pore Geometry 8, 25, 80, 85, 102, 126
Pore Radius 23, 42, 44
Pore Volume 10, 148
Porosity 4, 7 ff., 150, 178 ff.
Porosity Factor 10
Present Day Value 168 ff.
Pressure Bomb 119 ff.
Pressure Build-Up Tests 121, 127 ff.
Pressure Decline Curve 134
Pressure Gradient 58, 59, 77, 79, 152
Pressure Measurements 119
Pressure-Volume Relationship 47
Primary Gas Cap 71
Probable Reserves 154 ff.
Production Control Measurements 143
Production Rate 121 ff., 130, 141, 144, 172
Productive Area 152 ff.
Productivity Index 121, 197
Profit 167 ff.
Profitability Analysis 168
Proven Reserves 154 ff.
Pseudo Reynold's Number 80
Pulse Testing 137 ff., 152
PVT Relationships 46, 107

R
Recovery Factor 70, 73 ff., 102 ff., 153 ff., 169
Relative Permeability 69, 74, 81 ff., 102, 123
Reserves 147 ff.
Reservoir 3
Reservoir Energy 68, 119
Reservoir Geometry 169
Reservoir Limit Test 142, 152, 164
Reservoir Pressure 58, 59, 107, 121, 171
Reservoir Rocks 3
Reservoir Temperature 58
Residual Oil Saturation 82, 84 ff., 97, 102 ff.
Resistivity Index 30
Retrograde Condensation 53
Reynold's Number 20, 79
Rock Compressibility 17, 93
Royalty 167

S
Salt Cavern 3
Saturation 32 ff., 38, 42, 150 ff.
Saturation Pressure 47
Secondary Gas Cap 69 ff.
Sedimentation Procedure 23
Semi-Steady State Flow 77, 135, 165
Shaly Sandstones 30
Shrinkage (Factor) 49
Shut-in Pressure 121, 125
Sieve Analysis 23
Silicoclastic Sediments 4
Skin Effect (Factor) 130, 131 ff., 140, 197
Slipping Constant 80
Slugometer 145
Soil Subsidence 14, 16, 17
Sonic Depth Finder 120, 123
Spacing 170 ff.
Specific Inner Surface Area 24, 26
Specific Productivity Index 123
Sperry Sun Pressure Recorder 120
Stabilized Production 71
Stable Replacement 88
Static Pressure 127 ff., 140
Steady State Flow 77, 123, 165
Steep Slope (P_c) 38, 42
Stock Tank Oil 49, 121, 153

Super Compressibility 53
Surface Tension 35, 85, 106
Surplus 167 ff.

T
Taxable Income 167
Taxes 167 ff.
Tail Production 171 ff.
Thermal Shrinkage 48, 49
Thickness 129, 149
Tortuosity 24
Transient State 78
Transition Zone 40, 74

U
Ultimate Recovery 170
Unit Operations 169
Unproven Reserves 157 ff.
Unsteady State Flow 78, 140

V
Vapour-Pressure Curve 51
Viscosity 19, 52, 55, 71, 94, 126, 129
Viscosity Ratio 87 ff., 101 ff.
Viscous Fingering 100 ff., 164
Volumetric Method 150 ff.
Volumetric Reservoir 68

W
Water Break Through 97 ff.
Water Cut 90, 101, 104, 123
Water Drive 72 ff., 79, 103 ff., 107, 109
Water Encroachment Factor 109
Water Influx 108 ff., 164
Waxmann-Smits Correlation 31
Well Logs 14
Wet Gas Reservoir 50, 54
Wettability 3, 34

X
X-Ray Diffraction Analysis 23, 183

Z
z-Factor 54, 108, 151, 187